Digital Keywords

Tom Boellstorff and Bill Maurer, series editors

This series presents innovative work that extends classic ethnographic methods and questions into areas of pressing interest in technology and economics. It explores the varied ways new technologies combine with older technologies and cultural understandings to shape novel forms of subjectivity, embodiment, knowledge, place, and community. By doing so, the series demonstrates the relevance of anthropological inquiry to emerging forms of digital culture in the broadest sense.

Sounding the Limits of Life: Essays in the Anthropology of Biology and Beyond by Stefan Helmreich with contributions from Sophia Roosth and Michele Friedner

Digital Keywords: A Vocabulary of Information Society and Culture edited by Benjamin Peters

Digital Keywords

A Vocabulary of Information Society and Culture

Edited by
Benjamin Peters

Princeton University Press
Princeton and Oxford

Published by Princeton University Press,
41 William Street, Princeton, New Jersey 08540
In the United Kingdom: Princeton University Press,
6 Oxford Street, Woodstock, Oxfordshire OX20 1TR
press.princeton.edu

Cover and interior design by Amanda Weiss

Library of Congress Cataloging-in-Publication Data

Names: Peters, Benjamin, 1980– editor.
Title: Digital keywords : a vocabulary of information society and culture / edited by Benjamin Peters.
Description: Princeton : Princeton University Press, [2016] | Series: Princeton studies in culture and technology | Includes bibliographical references and index.
Identifiers: LCCN 2015037156 | ISBN 9780691167336 (hardcover : alk. paper) | ISBN 9780691167343 (pbk. : alk. paper)
Subjects: LCSH: Computers and civilization. | Information society—Terminology. | Digital media—Terminology. | Culture—Terminology. | Sociolinguistics. | Vocabulary. | English language—Etymology. | English language—Glossaries, vocabularies, etc.
Classification: LCC QA76.9.C66 D544 2016 | DDC 303.48/34—dc23 LC record available at http://lccn.loc.gov/2015037156

British Library Cataloging-in-Publication Data is available

This book has been composed in Arial and Sabon Next LT Pro

Printed on acid-free paper. ∞

Printed in the United States of America

10 9 8 7 6 5 4 3 2 1

To our students, past and future
And to those who inspired this book
and cannot read it

Words are actions, and actions are also a kind of words.

—Ralph Waldo Emerson

Contents

Acknowledgments

All collaboratively produced works exceed the expertise of any single scholar, and I suspect this is especially true of this volume. I have relied heavily on the generosity of friends and colleagues in crafting it. Some of those friends include Rosemary Avance, Biella Coleman, Christopher Kelty, Tarleton Gillespie, John Peters, Ted Striphas, Jonathan Sterne, and Fred Turner, among other contributors, who offered extensive critical comments and support at many crucial moments in the project. My colleagues at the University of Tulsa, especially Mark Brewin, Joli Jensen, and John Coward, have afforded a welcoming intellectual environment in which to do this work; and I also thank my students who have helped me test run some chapters in this volume—and have drafted a few of their own. The work has also benefited from the responses of anonymous readers at the Press and readers of early drafts at the scholarly blog *Culture Digitally* at http://culturedigitally.org/digital-keywords/. My editor Fred Appel has encouraged and guided the hybrid digital-print publication process throughout. Many organizational supports have made this volume possible as well: without the general wizardry of Hope Forsyth, Barbara Walters, and Jan Reynolds, and without the generous support of the Digital Working Group, the Center for Global Education, the Social Science Interest Group, and the Oklahoma Center for the Humanities at the University of Tulsa, this volume would not have been possible. My thanks also go to the Departments of Communication, English, and Languages at the University of Tulsa for their support. Of course the greatest debt goes to Raymond Williams, whose work remains as indispensable as it is inspirational. His pen leaves me continuously humbled at the fertility and the force revealed in the ever-unfolding relationship between language and the world.

Introduction

Benjamin Peters

In the age of search, keywords increasingly organize teaching, research, and even thought itself. Imagine for a moment an online universe without keywords: search bars would stay blank, log-ins and passwords would go unentered, and indexes and databases would rest ever listless and unpopulated. Keywords encode and decode the language of modern life. They stand sentinel to the halls of knowledge and power.

This volume is a timely update and celebration of the keyword studies tradition launched by the Welsh cultural and literary critic Raymond Williams forty years ago in his 1976 classic *Keywords: A Vocabulary of Culture and Society*. It is also an invitation to all those interested in the current information society and culture, as well as a provocation to the broad set of disciplines and traditions that employ its vocabulary. Oriented toward delivering foundational points about the current information age, this volume gathers and mobilizes diverse scholarly perspectives to serve a common set of core questions: What does the language of the information age do? How does that talk matter—how does it move, shape, and affect ways of being in the current media environment? What sources of power does our current vocabulary hide and reveal about our digitally lit world?

This introductory essay first announces the purpose, intellectual context, and history of the project before summarizing and grouping each of the twenty-five keywords from our current lexicon for discussing society and culture in light of information technologies. Each keyword chapter summary is also grouped into at least one basic grammatical category: subjects, objects, verbs, and prepositions—or, in other words, actors, things, actions, and environments. A comment on a few emergent themes in the crucial work of words in the information age concludes the essay.

The Purpose of This Volume in Context

This volume is no conventional or dry reference. Rather its purpose is to accrue lively resources in the emerging and sometimes miscellaneous field of digital studies. We repeat what Williams wrote in the introduction to his 1983 edition of *Keywords*: "This is not a neutral review of meanings. It is an exploration of the vocabulary of a crucial area of social and cultural discussion, which has been inherited within precise historical and social conditions and which has to be made at once conscious and critical—subject to change as well as to continuity." The volume takes an interdisciplinary snapshot of the evolving lexicon employed in humanistic and social scientific approaches to digital technology, offering up a small treasure chest of insight from contributors engaging with anthropology, communication studies, cultural studies, digital humanities, history, media studies, philosophy, policy studies, political science, religious studies, rhetoric, science and technology studies, social informatics, and sociology. The approach is manifold: sometimes this means scrutinizing relatively recent terms to take root in English such as **algorithm**, **analog**, **digital**, **hacker**, **internet**, and **meme**. Sometimes this means probing how older terms take on new uses—such as the **cloud** in cloud computing, the **mirror** in database mirroring, and the **forum** in online forums. Sometimes this means charting subtler shifts as classic terms such as **community**, **culture**, **democracy**, **memory**, and **sharing** migrate online and into digital forms. Throughout, this volume seeks to understand the transformational work played by socially significant words in the current information age.

Keywords matter. For Williams, language was not a transparent window to the world; it was—and is—one of the key epistemological materials of which the world is made. Keywords are not only metaphorical keys for opening new and hidden intellectual worlds—sometimes to dusty closets, sometimes to stadiums full of opposing crowds, and most often to corridors connecting to other corridors—although they certainly do that. Nor do keywords only open pathways for working across local webs of meaning, historical contexts, and the bustle and pushback of a material world, although they do that too. Rather keywords matter in ways instantly obvious to anyone who has ever used a search engine, entered a password, researched a question, or used an index. Keywords are the lexical operators of the current information age, and we might call a vast array of information technologies today *terminological technologies*—media from dictionaries to train stations, to computers, to Siri interfaces: all *terminals* that function in the language of the user; without specific keywords, terminological technologies do not work.

Consider how real life responds differently to whether a given person is named a *terrorist* or an *activist*, a *fetus* or a *baby*, *frail* or *cute*. Every word says something about the society it occurs in, and in turn is colored and given currency by that society. Language is what Kenneth Burke calls, in a usefully infelicitous phrase, a "terministic screen" of reality.[1] Terminologies do routine work: every word that empowers action also screens what we can do in reality because reality has first limited how we can use words. We must examine the real work that terms do (otherwise, we must remain silent), and, since Williams wrote, keywords have begun to do new work. In the age of search, digital keywords are no less than the obligatory passage point through which the semiotic material of history organizes life, and users in turn organize digital records. They are the axis upon which knowledge turns, the building blocks of all kinds of worldly webs, not least the World Wide Web itself. The ways search terms both constrain and overwhelm the organization of knowledge and action begin to illustrate the sweeping technical and organizational forces unleashed by the current communication revolution. The chapters in this volume share a focus on fleshing out our understanding of those forces, and of how digital technologies and discourses about them influence information societies and cultures in a globalizing world. As the chapters suggest, the significance of those forces is anything but negligible or obvious.

Like Williams, the contributors to this volume are fundamentally interested in the process by which our ways of talking and writing change on their own terms and, in turn, change the world. The determination of this critical-historical volume is to grasp the work words do. Scholars will also immediately notice that this volume is no faithful reproduction or extension of the format of Williams's work: this is a deliberate choice. The chapters do not always follow either the brilliant etymological method or the Marxian critical approaches that characterize Williams's studies. Adopting a broader format than did Williams, these chapters gesture toward the more encompassing and sustainable keyword project he began by ensuring that, as he did, each scholar approaches the keyword of choice in whatever way best fits the term's current relevance. For example, classic terms such as *community*, *culture*, *forum*, and *memory* bear meditations in the longer tradition of Western thought, while the meteoric rise of keywords such as *algorithm* or *geek* receive the full-court press of science and technology studies and critical feminist approaches. Neither the twenty-five essays selected here nor the appendix listing well over two hundred candidate digital keywords can pretend to comprehensiveness. We suspect no collaborative keywords project ever could—even one that benefits from both the flexibility of online interaction and the patient pace of print. In all, the resulting variety of approaches seeks to be loyal to the main point of Williams's critical-historical reference: each word that changes us deserves critical examination.

Williams began his own four-decade-long keyword study with a remark he made upon returning to Cambridge University after serving in World War II. He and another war veteran on campus observed about their nonveteran colleagues at that major and vibrant university, "The fact is, they just don't speak the same language."[2] During the war, while he was commanding tanks and witnessing the invasion of Normandy, something at the university had shifted: the underlying values and evaluations, "the formation and distribution of energies and interests," the discourse animating this "large and active university" were different. The search to understand why language had changed compelled Williams to complete, eleven years later in 1956, his pathbreaking *Culture and Society*, which charts the development of *culture* as a keyword. Before this seminal book was published in 1958, however, his editors excised from it an appendix containing a number of words of interest that, over the course of several decades, slowly matured into the 109 entries in the *Keywords* volume published in 1976, which he expanded and reissued with an additional 21 entries in 1983.

Like Williams's books, the present volume is preoccupied with the vocabulary of culture and society since 1945. Indeed every emerging generation of scholars would do well to admit and catalog, with Williams, the strange surprises of everyday language. Keywords arise at constitutive moments in modern media history, their semantic shifts voicing larger alliances and alignments of discursive power across the state, civil society, law, religion, economics, culture, technology, and the natural world—which is to say, *reality*. At times keywords even make history: "The emergence of a keyword in public discourse—whether a newly coined word or an old word invested with new meaning—may prove to be an illuminating historical event," writes historian of technology Leo Marx; "such keywords often serve as markers, or chronological signposts, of subtle, virtually unremarked, yet ultimately far-reaching changes in culture and society."[3] Marx notes that the French thinker Alexis de Tocqueville, in *Democracy in America*, had to coin a new French word, *individualisme*, to give "novel expression to . . . a novel idea." Williams too, in his survey of British culture and society, found that industrial capitalism had colored the very keywords—*culture*, *class*, *industry*, *democracy*, *art*—he had set out to understand. Economy, culture, and society, once examined, disclosed a reflexive interdependence with the currency of the terms that described them—a phenomenon Marx observed in his seminal study of the emergence of the term and concept *technology*.[4] In light of the ongoing information technological changes, we who live bit-saturated lives may feel tempted to echo Williams: the fact is, we just don't speak the same language that we did even a few years ago.

This work situates itself in conversation with three signal volumes published in the last decade, the first of which explicitly follows the larger cultural studies tradition of *Keywords* begun by Williams forty years ago: namely, Bennett, Grossberg, and Morris's *New Keywords* (2005), as well as Fuller's *Software Studies: A Lexicon* (2008) and Mitchell and Hansen's *Critical Terms for Media Studies* (2010).[5] Tony Bennett, Lawrence Grossberg, and Meaghan Morris offer a breathtakingly broad update to Williams with 144 very short keyword entries of public importance—from *aesthetics* to *youth*—even as that same update, now a decade later, no longer feels exactly "new." This volume's conceptual focus on digital discourse, itself an open experiment in unfinished projects, both narrows and responds to the awe many of us experience in encountering the ever-unfinished keyword project begun by Williams and continued by Bennett, Grossberg, and Morris, among many others.[6] Thankfully other works have contributed in their own way. Matthew Fuller's *Software Studies: A Lexicon*, for example, takes a similar approach, offering 38 short essays on terms central to the recent subfield of critical software studies—from *algorithm* to *weird languages*. This volume has benefited directly and indirectly from this literature: it includes brief keyword overlaps with Williams (e.g., *community* and *democracy*), with Bennett et al. (*community*, *culture*, *democracy*, *information*, and *memory*), and with Fuller (*algorithm*, *analog*, *information*, and *memory*). At other times, Bennett and Fuller's common inclusion of keywords like *copy* has licensed the contributors here to venture into its neighboring keywords like *mirror* and *surrogate*. Mitchell and Hansen's influential essay collection too shares only *memory* and *information* entry overlaps with this volume, although its rich and critical orientation toward Kittlerian media aesthetics, technology, and society can be discerned in such essays in this volume as those on *analog*, *cloud*, and *digital*, among others.

Whatever else it is, the digital revolution is a revolution in language. Peripheral keys have been reclaimed for everyday use—for example, the "@" for email, the "/" of URLs, and the "#" (once a "pound" sign on rotary telephones and now the "hashtag" of Twitter); our language morphs with new corporate capitalizations and spelling combinations such as *Facebook*, *Flickr*, *WordPress*, *YouTube* or *(micro)blogs*, *crowdsourcing*, *mashups*, *webinars*, *wikis*, as well as under the linguistic pressures of texting, as illuminated so marvelously in the study by David Crystal.[7] Digital work is about the print culture tradition of doing things in and by words. The *Oxford English Dictionary*, a hardy perennial in the study of English words, illustrates just how relevant a dictionary approach is to a keyword study of digital discourse. Consider a few entries and subentries that, alongside hundreds of others, entered that venerable dictionary in 2014 and 2015: *Bitcoin*, *BYOD* (Bring Your Own Device), *citizen media*, *hashtag*, *interweb* (humorous term for the internet), *LIFO* (Last In, First Out, a computing process with resonances back to at least Mark 10:31), *single-serving site*, *tweeting* (and *retweet*), *ubiquitous computing*, *VPN* (virtual private network), *webisode*, and the curious coupling of *photobomb* and *selfie*.[8] There is much to observe about this. It is notable how many entries refer somehow to the self; still more striking is the speed at which change in technological discourse outpaces our capacity—even that of our most admired keepers of the English language—to record such changes. Moreover, the *OED* spurs the reader to wonder at the extraordinary capacity of *all* language, once inscribed in media, to overflow the bounds we set on it. Dictionaries and keyword registers are becoming more relevant, as we all are swept up in the torrents and eddies, the current unrest of linguistic, cultural, and social change. There is no surer site for experiencing the concentrated superabundance of the English language than the *OED* today.

Williams, who served in the "corps of signals" and artillery in World War II, was of course no stranger to the extraordinary power unleashed by the techniques for inscribing language and signals. Twentieth-century philosophy—in particular the linguistic turn to ordinary or ideal language philosophy, semiotics, linguistics, and cultural studies—shares in common with the current digital age a deep interest in how language works, and in how propositional reasoning and its tools script our ways of being in the world. If Heidegger was right in calling language the "house of being," the plumbing and wiring of the house of modern-day humans appear to be undergoing a technical reconstruction and digital update. Many other recent theorists argue that material-semiotic linkages between language, institutions, and technologies inscribe and shape our current cultural, social, and political lives; and our investigations of humans as linguistic animals must now account for how information technologies inscribe, circulate, and pulse through current culture and society.[9] It may not be a stretch too far to claim, for example, that the core insight behind Michel Foucault's discourse analysis, Friedrich Kittler's discourse network (*Aufschreibesystem*, more literally "inscription systems"), and Bruno Latour's actor-network theory is the mutual inscription of material and semiotic power, of technology and language, in modern terminological techniques (archives, discourse, networks, etc.).[10] Language, once inscribed in technology and culture, reveals propositional forces known variously as data, information, and knowledge that continually remake our social world.

The fundamental question clamoring for attention in the background of all language analysis is this: How does language condition the ways we can be in the world? Perhaps nowhere is this more obvious than in the making of a keyword. A keyword is a socially significant word that does socially significant work. We seek to set apart a keyword primarily by what it does, not by what it means. Its work does not depend on its definition. It depends on how its uses separate and privilege certain practices, institutions, cultural norms, and doctrines over others. Cryptographers, programmers, and linguists, among others, refer to keywords as highly specialized terms designating identifiers that carry special weight in their analysis: the working definition here simply holds that keywords are those terms that do some heavy lifting or distinguishably significant work for analysis—and the work of each chapter here, then, is to spell out what that work is for each keyword.

If a keyword is a socially significant word, then what exactly is a *digital* keyword? In the most mundane and commonplace sense, a digital keyword is a keyword that refers to the recent rise in digital information technologies. Most chapters, for example, chart how their keyword "goes digital" in response to this straightforward question: What difference does it make whether or not a keyword refers to a currently bit-bathed world?

We may consider still another more basic and broader proposition about what makes a keyword digital: perhaps all keywords have always already been digital. A keyword is key only if the work it does can be distinguished from and then connected to that of other terms—a keyword must serve as a discrete operator in a larger semantic system. In order to become a term that bears distinguishable special weight in analysis, a keyword must first be fixed as a pivot point in a signifying system of discrete signals and meaning. In order to become a keyword, a word must first be subjected to some form of "digital" or discrete operations and codification. Google sells, for example, ad keywords such as *Insurance*, *Loans*, and *Mortgage* for top dollar because those letter combinations do countable work in the semantic markets of the financial industries, while, say, their anagrams (e.g., *I Care Nuns*, *Salon*, and *Gag Metro*) do not. Read broadly, the history of precise keywords stretches back as long as modern humans have manually used inscription tools to specify, index, and manipulate culture and society with language. (The **digital** chapter takes seriously what *manually* is doing in that sentence.) Semiotic precision is at least as old as language (e.g., the difference between, say, *bus* and *buzz* is fundamental to the intelligibility of both speech and writing), and standard spelling follows the development of nationalism and printing presses; but, as is well known to anyone who has tried to spell with autocorrect on, to do any programming, or to use alphabetized dictionaries, indexes, or catalogs, digital keywords propose—and fall short of—being about the disciplining work of saving ourselves with our own tools. We use digital keywords today to brand, mark, and clutch after the life rafts that buoy us in the present information deluge.

Keywords do not only organize the world for us. They also organize us in the world. For example, the terms I enter into a Google search in turn inform those searching for my user profile. Every search both consumes stored information and creates new information. "Digital keywords," as a simple search will suggest, are the provenance first of marketers and ad agents, and second of the information system that targets us with ads following our own past keywords. Keywords online are organized algorithmically, and not organically (see **algorithm**): invested interests and actors organize keywords. Not satisfied with the obvious examples such as search engine optimization (the industry practice of tweaking a site to optimize its visibility to search engine algorithms and subsequently increase visitor traffic to the site), this volume seeks to begin to document the larger technological, social, and conceptual forces that have combined to capitalize on the most recent chapter of terminological technologies.

The critical work of keywords in world history is of course greater than this volume can describe. It is a mundane yet profound observation that to speak is to act, and to inscribe in writing is an act of potentially enduring power. Keywords perform propositional forces in reality. For example, the names parents give their children, once recorded, outlast the living; set theorists invoke new possibilities with the scribble of a pencil; and again every online search is an act of both information consumption and metadata creation. Media history since at least the Bronze Age tells a longer story of the social power and modernization wrought by keyword technologies: lists and legal codes helped index and conscript empires, tax and domesticate early civilizations.[11] The movable-type printing press helped fix discrete keywords en masse in early modern Europe, and dictionaries and a slow flood of print material swept in the standardized spellings and canons behind the literate revelations and knowledge revolutions of the Protestant Reformation, the Renaissance, and the Scientific Revolution.[12] Control and communication technologies in the age of industrialization have also afforded the subsequent bureaucratization, statisticalization, and globalization of nationalist, regulatory, corporate, and academic knowledge capture, among other techniques for commanding the explosive forces of our industrial and information age.[13]

Since about the time Williams returned from the war, the English-speaking world has begun talking in the lexicon of information science: to choose one letter from that lexicon, we now find ourselves capturing, cataloging, categorizing, censoring, classifying, collecting, communicating, computing, and cultivating information (see **information**).[14] This sprawling expression of power can be grounded with a routine example: the passport may be understood as the keyword list—name, ID number, nationality, place of origin, sex, ethnicity, photo, a signature, and so forth—by which an authorizing state governs its citizens. A passport identifies a person with the very keywords over which the person identified has virtually no say. In both law and content, our identity documents are not our own; they point to others before and beyond us. The signature itself is a fascinatingly manual biometric technique for inscribing back into a larger register of citizens one's own identity: the scrawl of a signature, like a onetime pad in cryptography, seeks to be importantly *both* repeatable *and* inimitable.[15] The exercise of modern knowledge and power pivots on fixed terms.

How This Volume Came to Be

I first encountered Williams on the bookshelves in my childhood home and later in graduate school in the work of James Carey; a few years later, in the fall of 2012, my colleague Mark Brewin and

I found ourselves brainstorming a short list of words we found interesting. The project at hand quickly took wing and then molted from its initial conception about keywords in translation to a tighter focus on digital keywords: in the fall of 2013, an organizing committee of the Digital Working Group, which I chaired, invited a small gathering of scholars to draft short provocations on keywords of their choice at the scholarly blog *Culture Digitally* (culturedigitally.org), which were posted online for public review during the summer of 2014. Contributors were then invited to participate in a long weekend workshop held in the Zarrow Arts downtown facility of the University of Tulsa in Tulsa, Oklahoma, on October 10–12, 2014.

The rules of engagement at the workshop were simple if somewhat unusual: no papers were read out loud, since all texts had been circulated that previous summer. In fact the authors were invited to stay quiet about their own work. Instead, keyword panels featured a series of four or five prepared critical respondents—each of whom constructively critiqued one of the drafts posted online. Other scholars in attendance then triangulated and synthesized their comments across the feedback. This method helped optimize constructive revision suggestions as well as minimize defensive posturing (by my account, less than one minute in the long weekend went to publicly defending previous draft decisions). The dedicated discussion time also helped enrich the bigger-picture discussion about the critical and synthetic themes that introduce and interweave these chapters. These themes were rearticulated during a final internal round of peer review among the contributors in the fall of 2014, and then through the comments of outside readers at Princeton University Press.

The contributors have been selected in order to balance disciplinary coherence with interdisciplinary and international insight: generally media and communication scholarship, with supporting emphases in sociology, anthropology, and digital humanities, has the floor in this volume. Contributors were chosen in part for their willingness to make bold statements about digitally mediated culture and society that would appeal to more traditional areas of scholarly inquiry, such as the critical study of the economy and the environment, anthropology and religion, literature and philosophy. The same should be said for the intentional, however limited, international orientation of the volume: about a quarter of the contributors live in or hail from outside North America. While no single volume could ever represent all the key international and interdisciplinary aspects of the vocabulary of the information age, it is our collective hope that the conversations begun here will lead to more ecumenical discussion on information culture and society in a globalizing world.

The contributors are for the most part not specialists in language, lexicography, and etymology, even though some chapters deliver healthful doses of those approaches and no chapter goes without any reference to the word's history. For Williams, as for these contributors, language remains the vehicle, but not the end, of critical-historical analysis. The contributors selected these keywords not because their etymological record is necessarily the richest but because the core concepts are tectonic to the intellectual interests of these contributors. Sometimes the keywords name familiar areas of scholarly expertise about the digital age: Gabriella Coleman, John Durham Peters, Limor Shifman, and Thomas Streeter, for example, have recently published books on hackers, clouds, memes, and internet, respectively. More often chosen keywords mark areas of emerging expertise and fascination—Tarleton Gillespie's algorithm, Guobin Yang's activism, my own digital, Saugata Bhaduri's gaming, Christina Dunbar-Hester's geek, Gabriella Coleman's hacker, Bernard Geoghegan's information, Stephanie Schulte's personalization, Nicholas A. John's sharing—and many other contributors have also already published and continue working on other substantial scholarly projects that take up their obviously digital keyword. Similarly, other chosen topics, such as Jonathan Sterne's analog, Julia Sonnevend's event, Sandra Braman's flow, Adam Fish's mirror, Fred Turner's prototype, and Jeffrey Drouin's surrogate, bring to attention important and often overlooked keywords that find resonances in the work of these scholars outside of conventional digital discourse. Still others, such as Katherine D. Harris's archive, Rosemary Avance's community, Ted Striphas's culture, Rasmus Kleis Nielsen's democracy, Steven Schrag's memory, and Christopher Kelty's participation, reclaim for the digital age iconic terms with deep roots in social and cultural analysis.

No one can escape keywords so deeply woven into the fabric of daily talk. Whatever our motivations we—as editor and contributors—have selected these keywords because we believe the world cannot proceed without them. We invite you to engage and to disagree. It is this ethic of critical inquiry we find most fruitful in Williams. Keyword analysis is bound to reward all those who take up Williams's unmistakable invitation to all readers: Which words do unavoidably significant work in your life and the world, and why?

Search Results: Keywords in Review

This volume is about language in ways that resemble how a search engine sorts its results: both search through the inexhaustible repertoire of human thought, select desired results according to variable metrics, and express the results through inscription operations that bind fast language and reality, keywords and the actions. This section tries to take a snapshot of our research results—or, put more broadly, our doubly embedded language and world—by arranging brief summaries of the twenty-five keyword chapters into four basic grammatical categories that organize the English language itself: subjects, objects, verbs, and prepositions (or relational words) function as actors, things, actions, and environments (or surrounds that structure relations; stay tuned for more about the relationship between prepositions and environments). These categories are often most useful in their breakdown: the fertility of language handily dismantles such intellectual scaffolding (meant for swift construction and easier removal); and, as many chapters make clear, every keyword comes preloaded with polyphonic potential—one word can bear many perspectives, and the work of a word often manifests itself as it migrates across our mental categories. For example, words denoting what we often think of as an object, such as *prototype* or *mirror*, may best be understood variously as a subject, an action, or an environment that structures the set of possible relations between subjects, objects, and actions. Conversely, conventional verbal nouns or actions, like *sharing* or *flow*, may best be understood not as actions but alternately as subjects, objects, and environments. In short, the incomplete organization of these chapters into the four sections—subjects, objects, verbs, and prepositional environments—helps backlight how keywords do more work than we may think.

Subjects

Perhaps only two of the six chapters noted here—**algorithm**, **geek**, **hacker**, **meme**, **prototype**, **surrogate**—sit comfortably with the designation *subject*: **geeks** and **hackers** constitute two central (and significantly misunderstood) classes of human subjects that make up the technical expertise powering the current information age. The other chapters propose entities that, once made visible in a network of actors, reveal themselves as significant subjects hard at work in the modern media environment. New subjects emerge as the actions of conventional objects are viewed in context: thus in these chapters we see **algorithms** organizing programming and corporate discourse, **memes** migrating and multiplying online, **prototypes** prophesying in Silicon Valley, and **surrogates** populating the spaces between digital and print culture. These keywords act with enough force to belie their conventional designation as "just objects." Once these institutional, technological, and political networks have been mapped, **algorithms**, **memes**, **prototypes**, and **surrogates** join **geeks** and **hackers** as actors on the center stage of the drama that is digitally mediated behavior.

According to Christina Dunbar-Hester, the term **geek** has undergone a profound transformation in the age of computing: now detached from its pejorative association with circus freaks, no longer implying physical feebleness and weakness, the label today often applies to white middle-class males known for their technical expertise. This current use, Dunbar-Hester shows, underscores the need to situate the role that the technical classes play in propping up the global digital age. For example, while women are more likely to be computer scientists in Malaysia than in the United States, they are not necessarily more likely to be "geeks"; gender and technical affinity intersect with nationality and class in complex ways. Gabriella Coleman, too, critiques the stereotype of a **hacker** as a white male libertarian. In its place, and through a rich history of its varied sources and expressions, she uncovers an underlying hacker commitment to what she calls "craft autonomy," or the freedom to do technical work that motivates contemporary classes of computing experts. Hackers are not as we may have thought.

Four nonhuman actors (or at least they are not necessarily human)—**algorithm**, **meme**, **prototype**, and **surrogate**—announce themselves as subjects calling for attention. Tarleton Gillespie demystifies the many uses of the term **algorithm**, on loan from Arabic. It is at once a trick of the trade for software programmers, a synecdoche standing in for entire informational systems and their stakeholders in popular discourse, a talisman used by those stakeholders for evoking cultural authority and avoiding blame (e.g., to blame "Facebook's algorithm" implicitly shifts responsibility away from the company that designed it), and shorthand for the broader sociocultural shift toward, as Gillespie argues, "the insertion of procedure into human knowledge and social experience." In Limor Shifman's chapter on another commonly misunderstand term for discussing online culture, she offers a correction to the memorable myth that the internet is made of cats; rather, she insists, it is made of **memes**. In particular, she examines how the term *meme*, despite scholarly opposition and thanks to shifts in how users consolidate and share content online, has partly come to mean for internet culture today what a gene means for biology—namely, the smallest unit of transmission and variable reproduction. **Algorithms** and **memes** take up new forms of social life online, however purely technical these subjects appear at first glance.

Other technical subjects—**prototype** and **surrogate**—straddle and rework the divide between virtual and real, projective and past. Fred Turner submits as a new subject in the information age the **prototype**, or a working model that "make[s] a possible future visible." His analysis ties the Silicon Valley preoccupation with prototypes back to Puritan theology, showing how both cultures see in prototypes the foreshadowing of a brighter future (*typology* in Christian theology means the predictive interpretation of types and symbols binding past and future). By grounding, criticizing, and historicizing both the theology and the hucksterism at work in the term, Turner demonstrates how *prototype* points backwards in practice even as it professes to point forward to a model technology ready to symbolically save us from ourselves. As with Turner's **prototype**,

Jeffrey Drouin's analysis of **surrogate** complicates how a digital subject stands in—or serves as deputy or effigy—for the (print) object it reproduces. Drouin concretizes his Benjaminian analysis of how digital culture is at odds with print culture in a two-page spread of a Vorticist manifesto digitally reproduced online. Not simply does the digital version fetishize the "original" (a historical precedent set by print culture, which offers many copies and no easy originals, centuries before the coining of "the digital"), it also bears revolutionary uses. Instead of thinking how digital copy merely reduces the objects, he seeks to chart new relationships digital copies take up as they play surrogate to source materials. Once examined, keyword subjects such as **geek** and **hacker**, **algorithm** and **meme**, **prototype** and **surrogate** disclose previously hidden work.

Objects

Subjects act while objects are acted upon, or at least one grouping of the chapters below—**archive**, **cloud**, **information**, **internet**, **memory**, and **mirror**—may draw its mandate from that classic and contentious distinction in Western thought. Of course modern discourse, especially digital technology talk, traverses fashionably complex actor-networks and object-orientations that admit no such straightforward distinctions.[16] Every subject is also subjected, and every object acts upon us as we turn to it. Here too these lively objects do not remain mere objects for long. The **cloud** and the **mirror** stand in for metaphors for remote computing (or cloud computing and data mirroring), and in the process introduce new powers for communicating across distance and time online. **Archives** and **memory** in turn dig deep into the social construction and contestation of identity and meaning, and **internet** and **information**, by not always meaning what we think, too become, like the other objects, subjects in their own right and analytic lenses for focusing on the underlying actor entanglements. These object keywords, once analyzed, do far more than clarify philosophical muddling; they reveal powerful technological and institutional forces hard at work in the background of our analysis.

John Durham Peters unpacks the **cloud** in cloud computing—or the storing of data on remote servers. In his rich history of how clouds have long been at the forefront of science and imagination from ancient religion to meteorology and fractal geometry, among others, he argues that clouds today must be comprehended as anything but "purely immaterial, natural, and meaningless things." The current buzz about carbon-hungry "cloud computing" is neither natural nor environmentally friendly. Instead the cloud metaphor reveals how natural environments have long contained media, like clouds, that signal and structure, transform and elude our worldviews. Adam Fish too examines that perennial metaphor for reflection on the intertwined nature of the observing subject and the observed object—namely, the **mirror**, in digital mirroring or storing files at remote sites. Digital mirrors, for Fish, are sites of action for capturing, duplicating, and making visible information politics. His broad analysis spells out how mirrors have long replicated and distorted images, especially in the power differentials among cyberactivists and cloud computing companies.

Katherine D. Harris reflects on how **archives**, long understood as sites of copying and storing culture, become a potent site of differentiating between print and digital culture. Given that the verb *to archive* has long meant to reinterpret and canonize records, *to archive digitally* admits into play multiple interpretations of competing canons. The work of digital archives then can best be understood in light of the archivist, the database architect, the interface design, the uncertain sustainability of digital infrastructure, the act of reading, and the user experience, among other actors that make social both the texts and contexts of digital archives. Steven Schrag's treatment of that timely and timeless keyword **memory** charts several issues underlying these and other keywords concerned with digital culture. For Schrag, memory performs a curious balancing act between the material of "natural" memory and the technologies of storage media (between neural synapses and remote servers). Memory, digital and embodied, is central to our identities even as it extends beyond ourselves. Memory media render the past at once indelible, remixable, and riddled with gaps that seem to manifest themselves at ever-greater scales and speeds, although even that observation may be an artifact of imperfect recall. If **archive**, **cloud**, and **mirror** offer operations for making and using memory media, then Schrag returns us to the basic questions: Who or what remembers what, and how? Who controls our collective memory, and why?

Strictly speaking, there is no such thing as the **internet**. Instead, as Thomas Streeter shows, there are many competing categories for apprehending that term: among them, "hardware, software, protocols, institutional arrangements, practices, and social values." The term *internet* has "an outsize gravitational force"—it describes too much, marshals particularly modern (with a particular 1990s flavor) networking hopes and fears, and thus ends up meaning not nearly enough. Bernard Geoghegan, too, offers insight into the changing aspirations, institutions, and social practices of the information age through a keyword study of **information**. *Information* today means something very different from its medieval sense of that which gives matter form. He traces the modern technical sense of *information* as statistical measurement of serially patterned, nonanthropic traces to the nineteenth-century introduction of the electric telegraph and its instruments, standards, and economies. That new technical meaning, while narrow in itself, has gained huge purchase: consider how, for example, the object keywords **archive**, **cloud**, **internet**, **memory**, and **mirror** appear as subject sites insofar as they help us process **information**. *Information* today remains perhaps *the* seminal object toward which the modern digital age is oriented.

Actions

Underlying the difference between subjects and objects—whether an actor acts or is acted upon—is of course action itself. Perhaps all words must, in the end, be grasped in terms of actions they support and carry out. The action chapters that follow—**analog**, **digital**, **flow**, **gaming**, **participation**, **personalization**, **sharing**—explore digital keywords by critically studying the actions those terms imply.

The net publishes and privatizes the same data simultaneously, and our language has been adjusting to reflect that curious fact. **Personalization** and **sharing** appear to offer opposite updates for how digital media either narrow (*personalize*) or broaden (*share*) our reach online, but, upon closer investigation, both reveal corporations profiting off freely given user data. Stephanie Schulte diagnoses how **personalization**—or the technological targeting of individual interests and information by data service companies—appears to serve the liberal values of autonomous individuals while also enriching data companies. Personalization may occasionally deliver on its promises: it "connects users to one another . . . democratizes information, enables entrepreneurialism and civic engagement"; at the same time, Schulte warns, it also "commercializes culture and politics, alleviates productive discomfort, facilitates surveillance, and resituates or eradicates forms of agency." **Sharing** appears at first glance the perfect antidote to the privatization implicit in personalization, for sharing promises to collectivize. After all, what is sharing about if not community and, according to the aphorism, caring? *Sharing*—derived from the same root as *shear* or to divide up, as in a *share*holder, but also with a major role in twentieth-century therapeutic discourse—now promises to enrich the social life of Web 2.0 users as well as the pockets of large data companies that profit by selling, not sharing, freely shared user information. Let us not forget: the online user, whether **personalizing** or **sharing** data online, is also always used.

Flow and **gaming** frame the forms of collective social action and connection that have long taken place offline, and now speak especially powerfully to our networked systems. Sandra Braman observes that, for all the talk about the electronics and logics, media scholars in the information age are curiously preoccupied with a hardy metaphor of **flow**. Referring to matters that go far beyond the broadcast media content to which Williams applied the term in his breakthrough book *Television: Technology and Cultural Form* (1974), the concept of flow is used today to think about what happens in technical, social, and sociotechnical systems through which human-human and human-computer interactions unfold at the individual, group, and societal levels. Whatever the kind of system we are talking about, it is flow that makes it possible. For Saugata Bhaduri, **gaming**, like flow, is about continuous action, but unlike flow, that action comes with a distinct sense of social risk. Although his analysis only begins to hint at who bears the risks of online gaming today, his analysis of *gaming* as risky collective action—the word is derived from the proto-Germanic sense of "people together," coupled with the English suffix for the present continuous—reveals how the term and its variants (e.g., *gaming a system*, *gamesmanship*, *gamification*, *gambling*) infuse online gaming subcultures (hacks, mods, cheats, sandboxing, fanfiction, cosplay, machinima, etc.) with an older sense of subversion. These action keywords—**flow**, **gaming**, **personalization**, **sharing**—state something very basic about all keywords: our language matters most for what it does, not what it means, and the social risks and tradeoffs built into these actions are sweeping our digital environments. Christopher Kelty makes this underlying point clear: in mining the ever-present yet overlooked intellectual roots of **participation** as both a word and a concept in political thought, he reveals that, as with perhaps all action keywords, to participate is to belong collectively, although not always voluntarily.

Now consider two framing keywords for the project. In information-age talk, **analog** and **digital** usually appear as both inseparable and opposed—as if they were two bits in a binary relationship, off and on, 0 and 1. On the contrary, an organizing point of this volume holds that digital and analog categories are *not* binary opposites: the digital is not synonymous with only artificial, discrete, finite symbols, nor is the analog identical to all that is *not* digital and to all that *is* natural, continuous, real waves. Rather the key to grasping that **analog** and **digital** are not reciprocals begins with an analysis of the actions, not the forms, these terms bear out. In the manifold openings and nonrelations occupying these two keywords, we uncover fresh insight into many current misunderstandings and themes animating the information age.

Specifically, Jonathan Sterne argues that the **analog** is not the opposite of the digital. He traces two tracks the term has followed since the 1800s: first, *analog* is employed when a specific technical process is used to represent another (analog computers that use voltage or water). He notes that the process has no necessary relation or opposition to discrete digital computing. Second, crystallizing in the writing of Stewart Brand in the 1980s, *analog* took on an expanded and misleading denotation as the negative of digital—or everything that is not digital, and thus all material reality. Only recently has our talk tried to subsume nature itself into *the analog* as a way of distinguishing it from the dawn of *the digital*.

Several problems follow this strong distinction: there is nothing natural about analog machines, nor are analog and digital techniques necessarily incompatible. In fact a lot of our so-called digital devices are analog, right down to the voltages on the logic board. The digital/analog binary tracks back to the binary thinking of digital theorists, not the binary nature of digital technologies—a false binary that Sterne unravels. In my chapter, I too seek to understand the term **digital** in its original sense—in terms of digits or *fingers* that count (compute), point (index), and manipulate; this bit of triadic thinking aims to help further tease apart the unhelpful pairing of digital-analog. The postwar explosion of computing power (in rough parallel with Claude Shannon's sense of information) has catapulted the counting and computational sense of *digital*, but not the others. *The digital*, if grasped only narrowly, will remain a quintessentially twentieth-century buzzword, even as its techniques continue to spread into the current century: "digital television," for example, now passes simply as *television*, and the apex of *the digital*—the notion of the convergence of all that is countable, or a digital singularity—now sounds quaintly late twentieth-century. In my brief speculative history from index fingers to file cabinets, *digits* appear among those media that have long indexed and manually manipulated many forms of information society and culture: indeed much of this work is devoted to demonstrating how digital environments manipulate how we talk and who we are.

I have grouped *analog* and *digital* here in the action keywords section in order to emphasize how they, like actions, coexist without suffering from the loggerhead logic that jams up our thinking about ontological states of being and categorical forms. Actions can happen simultaneously in the same space: for example, the work of **digits** and **analogs** coexists just as easily as one can, say, **share flows** of **participation information** by **personalizing** one's **gaming**. As these and other chapters suggest, perhaps the most specific lesson to take from keywords is what actions they commit (analog represents and waves; digits count, point, and manipulate; etc.). What a keyword does is both more relevant and more interesting than what it is, and keywords are among our many linguistic analogs for describing our active world.

Environments

Prior to subjects (actors), objects (things), and verbs (actions) are the infrastructural surroundings and grammatical conditions in which reality and language operate. Both prepositions and environments disclose the hidden infrastructural surrounds that shape the set of possible relations, whether the relationship between words in language (grammar) or the relationship between actors in reality (environment). Both grammar and environments also signal their inner workings with the subtlest and smallest of signs—and prepositions are among such potent forms (punctuation, formatting, and design are others). Prepositions are relational words such as *of* and *for*, and spatial-temporal relations such as *in*, *under*, and *before* that, like environments, organize the relations between subjects, objects, and actions. Their work, like the room or medium in which you are reading this book, is to go unnoticed. (If I asked how many prepositions are in this sentence, most readers would have to stop and reread it.) To borrow a phrase originally about infrastructure from Geoffrey Bowker and Susan Leigh Star good environments are hard to find.[17] They are harder still to fashion, since environments invariably fashion us first, even though the current ecological crisis underscores just how transactional the modern human relationship with the world is. It is no surprise that a spate of recent scholarly attention has poured onto digital environments—applications, architecture, grammars, infrastructure, platforms, scenes, settings, standards, structures, and (operating) systems make up all the hard stuff that usually goes invisible. Like a preposition in a sentence, an environment lies outside what we readily sense and read. It shows itself in the cracks.

Consider, first, two keyword chapters—**democracy** and **activism**—that concern not so much environments themselves as the everyday practices and ambiguities constituting the environments for political action. (By contrast, keyword chapters on **community**, **culture**, **event**, and **forum** speak directly to the larger conceptual environments that structure past, present, and future information ages.) If **democracy** is a universal aspiration, Rasmus Kleis Nielsen intimates, then digital democracy may be among its most cherished slogans. This chapter punctures that slogan by observing that digital technologies have in fact had modest, indirect, and internal effects on the functioning of democratic institutions, but not at all in the deliberative, direct, and participatory ways theorists identify as common to both digital technologies and liberal representative democracies. The digital and the democratic do not necessarily have anything to do with one another. To imagine only what could be, and to ignore what actually already is, democratic, Nielsen concludes, dulls our ability to assess and address real democratic practice and problems. Guobin Yang takes a different critical approach. He argues that by identifying the ambiguities and biases in current discourses about online **activism**, one may resuscitate and reaffirm the meaning and politics of protest—which is in fact the new normal: all street protest today takes place with some form of digital organization or documentation. His central premise is that the proliferation of ways of talking about activism has weakened it as a political practice, and he illustrates this in discussing analytic ambiguities besetting modern protest in the West and in China. Activism discourse in China, for example, remains ambiguous, since protest can both subvert and stabilize (by making visible and thus more resilient) authoritarian state oversight. He argues that these ambiguities underscore the docile elements of digitally connected protest and threaten to undercut the potential radical power of activism. Together Nielsen and Yang introduce how discursive environments—namely, modern democratic society—shape and in turn are shaped by (political) action.

Reflections also follow on how several classic environments—*community*, *forum*, and *culture*—have gone digital: Rosemary Avance paints a rich intellectual backdrop for the most frequent of these twenty-five keywords in the English language—**community**. Her treatment of scholarship from the field of religion and media seeks a middle way between puncturing the utopian rhetoric of online communities and embracing the genuine experience of belonging to something larger than oneself. With the caution "not every site that calls itself a community is one—just as not every site that does not, is not," her analysis clarifies digital, virtual, and hybrid gathering distinctions, as well as collapses the conventional online-offline divide. Like community, a **forum**, in Hope Forsyth's analysis, is also a liminal environment for gathering social life around shared civic, commercial, religious, and legal interests. Forsyth posits how forums require "human-supporting infrastructure" to meet the physical necessities of its visitors. While the internet alone cannot supply such infrastructure, once combined with physically sustaining spaces—such as the coffee shop, be it an eighteenth-century European café or a modern Wi-Fi hot spot—forums act as embodied physical environments for fostering societal interaction.

Culture is the keyword among keywords for Williams (who contributed to the founding of cultural studies between the 1960s and the 1970s). It is probably also the archetype of discursive environments. One of his careful readers, Ted Striphas, offers a sensitive update to Williams and a wide-ranging intellectual history, describing how culture has coevolved with the digital turn since the end of World War II. No longer the antithesis to technology, culture has recently interpenetrated with the computational (e.g., digital humanities, culturomics, and big-data-driven cultural studies). The current state of culture is a testament to the "dynamism and adaptability of [what Williams calls] 'one of the two or three most complicated words in the English language.'" Finally, we encounter another chapter that seeks to frame the whole digital age: an **event**, according to Julia Sonnevend, is an *important happening* or occurrence inscribed into history. Offering a theoretical framework for describing the extended process through which occurrences "eventually" become events, her goal is not simply to understand digital media events, such as the death of Steve Jobs, or even general media events such as the collapse of the Berlin Wall. Rather she describes the process through which all events come to be as that of actors founding, universalizing, condensing, counternarrating, and then diffusing narratives across borders. The digital age will one day become an event in history—and future historians will see the vocabulary of our information society and culture as an event itself. This volume seeks to accrue intellectual resources to help eventuate that end.

Emergent Themes and Concluding Comments

To risk overprecision, the total number of connections between n nodes in a network can be expressed as the number $n(n-1)/2$ (which is also a triangular number equal to the sum of every positive integer one less than the number of nodes). In other words, with twenty-five chapters, there are $((25 \times 24)/2 = 24 + 23 + 22 \ldots + 1 =)$ 300 possible connections between keywords in this volume alone. I limit myself here to three basic orientations and a couple of concluding comments. Those seeking more connections are invited to reflect on the cross-referencing *See in this volume . . . See in Williams* sections found at chapter ends that link individual keywords to other keywords in this volume and in Williams.

Perhaps three operations can be said to describe how these essays process our social and cultural vocabulary: discursive subtraction (critique), addition (reclamation), and multiplication (complication), although we leave the crucial task of division (critical analysis) to the reader. Without question, the most common orientation of chapter arguments in this volume is critical or subtractive. A chorus of voices—including those in **activism**, **cloud**, **community**, **democracy**, **digital**, **internet**, **mirror**, **personalization**, **prototype**, **sharing**—declares that much talk about the current information age is bunk. We do not buy it. We'd be better off without it. A certain subset highlights what happens when aspirational ideals break against the rocks of messy embodied practice: for example, when **community** falters in person as well as online, when tech-savvy **democratic** campaigns still canvas the streets and knock on doors, and when the heated debates of online chat **forums** suddenly need a bathroom break. Keywords—it should surprise no one who has walked the stacks of a library or peeked beyond the first page of search results—produce mounds of misleading hits. Given a googol search results, all but a few must be garbage to the user. There is much to subtract in a world organized by keywords.

No other theme in this volume rings out as clearly as the call to identify and hold responsible those whom digital discourse serves, although the summary critique goes beyond finger-pointing to call for the struggle for social change. Questions abound: How are interested actors and institutions shaping and exploiting the current digital lot, and to what effect? For example, with whom are we sharing when we share online? Where is the cloud? Whom do we see in the (data) mirror? Whom do we serve while democratizing, personalizing, and prototyping our media and communication? How can we reflect, reclaim, and reform the ways modern language and its technologies serve our social lives? Under what conditions could a more equitable and beneficial world be brought about? In response, nearly all chapters press for more attention to the politics of the ideas and institutions steering and mediating everyday life. Chapters such as **community**, **event**, **forum**, **hacker**, **memory**, **mirror**, **prototype**, and **surrogate** point out potentially misleading aspects of the authenticity claims their keywords can make in a so-called virtual environment. The labels of **community**, **event**, and **forum** are often used to authenticate digital environments for what they are not, just as the analyses of **geeks**, **hackers**, and **memes** help contest and complicate once seemingly uniform actor classes.

The second orientation, then, is toward reclaiming some residual meaning in our terms and in the process revealing something significant that has been long hidden in plain sight. Among other chapters, **analog**, **cloud**, **culture**, **flow**, **forum**, **geek**, **hacker**, **meme**, **participation**, and **prototype** suggest new ways of thinking with language: **analog** has no necessary relationship to nature or reality, but it does have meaningful purchase on the techniques of representation; **culture**, by blending with information technology, renews its staying power as a significant frame for life; **hacker** has less to do with the politics of freedom itself than with the freedom to create and work in technologically constrained environments; **prototypes**, by projecting the future, invariably ground us in the ever-present history of questionable typological thinking. Many others could be listed as well.

The third orientation is toward complicating the uses of the keyword—or to add to analysis not just one but multiple distinct threads of meaning. Chapters such as **algorithm**, **analog**, **archive**, **digital**, **event**, **gaming**, **memory**, and **surrogate** leave the reader chewing on multiple meanings—sometimes countably many: **analog** has at least two pathways (representative and nondigital), **digital** has at least three (counting, pointing, manipulating), **algorithm** comes in at least four guises (trick, synecdoche, talisman, procedure), and **events** unfold in a five-step process (found, universalize, condense, counternarrate, and diffuse narratives). None of these chapters are complete or conclusive: rather discursive multiplication helps tease apart the multiple uses—the warp and the weft—tightly woven into the fabric of our language.

Several overarching themes emerge from the chapters as well. If there is a consensus position among the chapters (a fiction imagined by many editors), it is the definitive lesson that all forms of media stand in imperfectly for other forms of media. Consider print culture: before the e-tablet ever stood in for the paperback, the paperback stood in for the hardback, and the hardback for the manuscript, the manuscript for the codex, the codex for the scroll, and the scroll for the e-tablet's namesake, the clay tablet of antiquity. Digital **communities**, **mirroring**, and online **persona**, for example, all derive their metaphoric power from claiming to stand in for a supposedly more substantively real entity elsewhere—an organic community, the source of the mirrored image (or file), and the living person herself. Digital discourse is the new kid on the old block: as **cloud**, **community**, **memory**, **mirror**, **prototype**, and **surrogate** propose, the search for what is original, authentic, and real in human life has proved elusive since well before those values were enshrined in the Enlightenment.

Our metaphors elude us in part because they fall into the blurry neutral zones connecting the supposed conceptual divides between the technological and the natural, the organizational and the organic. Language, once examined, muddles this conceptual sort of digital divide. Consider the natural sources of the technological metaphors in technical **analog** from biology, the **cloud** (computing) in the sky, (online) **communities** since settled human history, (online) **culture** since agriculture, the **digit(al)** on our hands and at our fingertips, (system) **flows** from rivers and streams, **memes** from the combination of genes and memory, (flash) **memory** from human memory, (file) **mirroring** from the early mirrors of polished obsidian stone, and (online) **personalization** of a person herself—each term speaks to a much broader worldview. The language of modern technology draws deep from the word wells of media history.

This metaphorical porting from one material state to another evades us for other reasons. It is not that the metaphors are wrong to cross categories (technological and natural, the symbolic and the material, the digital and the analog). It is that the categories have already always been mixed. This is Williams's point updated for the digital present: material and symbolic production in modern life must be understood together, not separately.[18] Material and symbolic production converge precisely in information society and culture. The technological and the natural are not philosophically incompatible categories; *analog* and *digital* are neither opposed nor fused. The virtual spaces we inhabit cannot be separated from the natural world our digital devices imitate, reproduce, and sap. In the big view of media history and philosophy, the digital is profoundly normal—and normal is profoundly fascinating and worth criticizing. Power has been concentrating unevenly since the end of the cosmic inflation—a tiny fraction of a second after the Big Bang (or, more precisely, sometime between 10^{-32} or 10^{-33} seconds later). If power imbalances may then be considered natural, then modern media and terminological technologies invite us to confront what is most "natural" for humans: as the long and industrious history of the creative universe and our inventive species *Homo sapiens sapiens* indicates, terminological technology is in our nature. We live by art and artifice alike. Careful reflection suggests that media rest on the deeper orthogonal overlap, not the conceptual separation, of natural and technological resources. Media are our lot, our environments.[19] Modern culture, as Hillel Schwartz details at length, emerges out of the comingling, not the categorical contrast, of so-called natural sources and their technologies that reproduce them so uncannily.[20] The craftsmanship of natural scientists, natural historians, and natural number theorists challenges traditional notions of "natural": their trades teach how intensely virtual, technical, and even *digital* nature can be.

Several keyword chapters speak to how media blur categorical states of time (past, present, future) as well. **Algorithms** function as talismans for larger institutional trajectories; **archives** present themselves as present versions of the past; **digits** index longer, more diverse media histories and less singular media futures; **events** inscribe the past onto the historical record and punctuate it; **gaming** and gamification management strategies signal chances to level up in the future; **prototypes** project a future that points to the past; and **surrogate** texts deputize the present encounter with past copies. In other words, **prototypes** converge moments of the present and the perceived future, just as **archives** do the same for the present and the perceived past. In each case, keywords show media to be prime time-axis manipulators—a political fact so bald it stares us in the face every time we gaze on the pause and replay buttons, audio and video progress bars, and text scrollbars at the edges of our screens.

In the work of history and time bending, religious thought also colors digital keywords by shades, as suggested in the **cloud**, **community**, **digital**, **forum**, and **prototype** essays, among others. Digital media do not represent, reproduce, or save our world across time and format, although they make imperfect efforts to do precisely that. Media language is irreparably superabundant in its theological overtones—it is beyond representation and at once powerfully inscribed into the saving techniques of modern life: its metonymies, proxies, surrogates, synecdoches, and analogs have long promised that modern devices might let us save ourselves by replicating versions of ourselves—something our species has been doing for generations, although the current digital and ecological crises threaten to condemn us in terms we might call "theotechnical." Consider the bureaucratic-theological mingling in how we talk about, for example, saving a file: to save a file is not simply to copy a file. It is to make content new by giving it a different name. To save a file is to save a proxy of itself under a new name and a different time stamp. Media talk brims with troubling salvific force.

To summarize a few of these themes, the keywords outlined in this volume continue the monumental task of discovering and confronting the power words wield in society and culture. Digital keywords, once studied, do very old things: they impinge on our analytic arithmetic for understanding the past, the present, and the future; they complicate our distinctions, natural, artificial, and human; and they reveal the adamantine institutional and intellectual forces thought to be scripting our lives, sometimes even afterlives, always hard at work in the present. As Williams wrote, "If the social is always past, in the sense that it is always formed, we have indeed to find new terms for the undeniable experience of the present: not only the temporal present, the realization of this and this instant, but the specificity of the present being, the inalienably physical, within which we may discern and acknowledge institutions, formations, positions, but not always as fixed products, defining products."[21] That specific present is now, and in the following analysis of the modern reality ushered in with these new terms, we must attend to our current linguistic lot. These essays draw out several themes from this small sample of the vocabulary of information society and culture, including Heideggerian reflection on the root relationship between language and practice; the productive tension between disciplinary specialization and generalist insight; critical attention to the religious and theological overtones of salvific media discourse, and to the institutions marshaling media-technical power, among others.

We must not imagine that digital keywords should (or somehow could ever) be stopped from drawing on natural and cultural resources for inspiration; by the same token, nor should we neglect the uses, consequences, and benefactors of that language: if left to the sloganeering of the unscrupulous and hucksters, the inexhaustible excess of digital keywords (among other media metaphors) will surely be channeled into covering up the significant consequences and costs of the profitable appropriation and exploitation of natural and cultural resources, both material and metaphorical. Consider a few examples: big data are not new because data are big. (Data set sizes have long been growing colossal, although digital computing marks a threshold in its exponential acceleration.) What is new about big data is the invasive inferential power of pinpoint granular data analysis, not the trivial scaling of data. The dark side of big data, in other words, is how scalably small analysis now is—its penetrating zoom. Similarly, "personalizing" your media, simply put, means that both you and others get to see more of yourself, although who the others are is not up to you (if ever it was). The "cloud" materializes in climate-controlled warehouses running stacks of data servers squirreled away from public view. Or the "media ecosystem" metaphor for the relationship between traditional news journalists and bloggers, for example, might sound like a healthful symbiosis and self-sustaining media environment, while in fact a glance at the hemorrhaging news industry calls to mind the industrial equivalent of natural selection.

This much is obvious: digital discourse demands active scrutiny. Every act of naming a keyword is an investment of institutional power—or, to paraphrase Hegel, naming is an act of sovereignty.

To name a keyword thus is always to raise questions about who names what and why. The chapters that follow instruct to this basic fact: the power of interested actors and institutions is inseparable from the language that exercises that power. Our language has long pivoted on the politics of institutions, social norms, practices, organized interests, and other cultural material forces at work in the terminological technologies that populate information society and culture (not just modern-day algorithms, archives, information, memes, networks, prototypes, and search, but also research, speech, and script themselves).

In the end, though, the superabundance of language leaves us with even more than the need for self-scrutiny between the competing forces of those who sell and those who think, however compelling that distinction may be: because keywords, once critically inquired after, continue to abound in potential uses, their moving power can and should be redirected and rechanneled in the service of the many, not just the few. With reflection and work, keyword analysis may do more still. It may effect an educational change, instilling and renewing a sense of awe at the wider world beyond even the enduring calculus of politics and competing centers of power.

I now invite readers to turn their critical attention to the following contributions of scholars who have privileged the marvelous medium of words in critiquing, revealing, complicating, and reforming our modern-day information vocabulary. To borrow and twist a phrase from Heidegger, the essence of digital keywords is neither the digit nor the keyword: once examined, these essays redirect literate attention from whatever the most recent terminological technology may be to perennial and pressing questions of the aesthetic, cultural, economic, ethical, historical, legal, medical, philosophical, poetic, political, religious, social, and much else. (Even this list, like the table of contents, is an act of alphabetic artifice.) They invite us, as all keywords ought to do, to reflect on the larger universe and the terms that position us in it.

Tulsa, OK
August 15, 2015

Notes

1 Kenneth Burke, *Language as Symbolic Action* (Cambridge: Cambridge University Press, 1966), 45.

2 Raymond Williams, *Keywords: A Vocabulary of Culture and Society*, 2nd ed. (London: Fontana Paperbacks, 1983), 9.

3 Leo Marx, "Technology: The Emergence of a Hazardous Concept," *Technology and Culture* 51(3) (July 2010): 562–63.

4 Ibid.

5 Tony Bennett, Lawrence Grossberg, and Meaghan Morris, eds., *New Keywords: A Revised Vocabulary of Culture and Society* (Malden, MA: Wiley-Blackwell, 2005); Matthew Fuller, *Software Studies: A Lexicon* (Cambridge, MA: MIT Press, 2008); and W.J.T. Mitchell and Mark B. N. Hansen, *Critical Terms for Media Studies* (Chicago: University of Chicago Press, 2010).

6 For other roughly related projects, see Reinhart Kosselleck's *Geschichtliche Grundbegriffe: Historisches Lexikon zur politisch-sozialen Sprache in Deutschland* (Stuttgart: E. Klett, 1972–97), the Raymond Williams Society journal *Key Words*, the *Theory, Culture & Society* New Encyclopaedia Project, and the Keywords Project at the University of Pittsburgh (http://keywords.pitt.edu/whatis.html), among others.

7 David Crystal, *Txting: The Gr8 Db8* (Oxford: Oxford University Press, 2008).

8 See a full list of previous updates, including new entries in 2014, here: http://public.oed.com/the-oed-today/recent-updates-to-the-oed/previous-updates/.

9 A long tradition in philosophy dating back at least to Spinoza has considered the almost metaphysical debt borne by reality to language. A few recent thinkers concerned about the linguistic turn and language analysis include J. L. Austin, Judith Butler, James Carey, Jacques Derrida, and Richard Rorty. (For more, see Richard Rorty, ed., *The Linguistic Turn: Recent Essays in Philosophical Method* [Chicago: University of Chicago Press, 1967].)

10 For more on "material-semiotic actors," see Donna Haraway, "The Promises of Monsters: A Regenerative Politics for Inappropriate/d Others," in *Cultural Studies*, ed. Lawrence Grossberg, Cary Nelson, and Paula Treichler (New York: Routledge, 1992), 295–337; Donna Haraway, "Situated Knowledges: The Science Question in Feminism and the Privilege of Partial Perspective" and "A Cyborg Manifesto: Science, Technology, and Socialist-Feminism in the Late Twentieth Century," in Donna Haraway, *Simians, Cyborgs, and Women: The Reinvention of Nature* (New York: Routledge, 1991), 149–81 and 183–201, respectively.

11 See Jack Goody, *The Domestication of the Savage Mind* (Cambridge: Cambridge University Press, 1977); Eric Havelock, *Preface to Plato* (Cambridge, MA: Harvard University Press, 1963).

12 See Elizabeth L. Eisenstein, *The Printing Press as an Agent of Change: Communications and Cultural Transformations in Early Modern Europe*, 2 vols. (Cambridge: Cambridge University Press, 1979); Bernard Siegert, *Passage des Digitalen: Zeichenpraktiken der neuzeitlichen Wissenschaften 1500–1900* (Berlin: Brinkmann Bose, 2003); Cornelia Vismann, *Files: Law and Media Technology* (Stanford, CA: Stanford University Press, 2008).

13 See Max Weber, *Economy and Society: An Outline of Interpretive Sociology* (first published in German in 1922), ed. Guenther Roth and Claus Wittich (Berkeley: University of California Press, 1968); James R. Beniger, *The Control Revolution: Technological and Economic Origins of the Information Society* (Cambridge, MA: Harvard University Press, 1986); Daniel R. Headrick, *When Information Came of Age: Technologies of Knowledge in the Age of Reason and Revolution, 1700–1850* (Oxford: Oxford University Press, 2000).

14 See the popular account by James Gleick, *The Information: A History, a Theory, a Flood* (New York: Knopf Doubleday Publishing Group, 2011).

15 Craig Robertson, *The Passport in America: The History of a Document* (New York: Oxford University Press, 2010).

16 For two examples of the emerging literature on actor-network theory and object-oriented ontology and philosophy, see Bruno Latour's *Reassembling the Social: An Introduction to Actor-Network Theory* (Oxford: Oxford University Press, 2007) and Graham Harman's *Tool-Being: Heidegger and the Metaphysics of Objects* (New York: Open Court, 2002). While the latter is intensely helpful in conceptualizing the meaningful connections between and in things, I should note that I find Harman's rejection of the linguistic turn mistaken or hasty, in part because, as corners of both language-oriented and object-oriented philosophy argue, there is a fundamental connection between the material-semiotic work of language as objective action and objects as linguistic actors.

17 Geoffrey C. Bowker and Susan Leigh Star, *Sorting Things Out: Classification and Its Consequences* (Cambridge, MA: MIT Press, 1999).

18 Williams, *Keywords*, s.v. "Culture."

19 John Durham Peters, *The Marvelous Clouds: Toward a Philosophy of Elemental Media* (Chicago: University of Chicago Press, 2015).

20 Hillel Schwartz, *The Culture of the Copy: Striking Likenesses, Unreasonable Facsimiles* (Cambridge, MA: MIT Press, 1996).

21 Raymond Williams, *Marxism and Literature* (New York: Oxford University Press, 1977), 128.

Digital Keywords

1

Activism

Guobin Yang

First published in 1976, Raymond Williams's *Keywords* captures the spirit of his times. The 110 entries in the first edition include *radical*, *revolution*, and *violence*. *Liberation* is one of the 21 entries added to the second edition in 1983. These words linked together the worldwide revolutionary movements of the 1960s and early 1970s.

From the vantage point of the twenty-first century, however, the absence of the word *activism* in Williams's classic is conspicuous. In the decades since its first publication, but especially since the 1990s, *activism* has become a popular word in contemporary cultural and political discourse. It is used not only by citizens and civil society organizations, but also by government bureaucracies, international agencies, and even business corporations. Furthermore, the growing popularity of *activism* is accompanied by the declining use of *revolution* and *liberation*, or at least declining up until the "Occupy" movement and the Arab Spring protests. What does the ascendance of *activism* reveal about contemporary culture, society, and politics?

An Ambiguous Word

Activism is an ambiguous word. It can mean both radical, revolutionary action and nonrevolutionary, community action; action in the service of the nation-state and in opposition to it. This ambiguity has existed since its first usages in the early twentieth century. The German philosopher Rudolf Eucken used the term in his 1907 book *The Fundamentals of a New Philosophy of Life* to refer to "the theory or belief that truth is arrived at through action or active striving after the spiritual life" (*OED*, 3rd ed.). In continental Europe during World War I, *activism* meant "advocacy of a policy of supporting Germany in the war; pro-German feeling or activity" (*OED*, 3rd ed.). The word in 1920 began to take on the more general meaning of "the policy of active participation or engagement in a particular sphere of activity; *spec.* the use of vigorous campaigning to bring about political or social change." By 1960 the term could have an almost incendiary connotation, as in "The sizzling flame of activism is visible in both the agricultural and pastoral districts" (*OED*, 3rd ed.). About *activism* as a service to the nation-state, Hoofd (2008) writes:

> Etymologically, 'activism' has strong affinities not only with an essentially transcendental philosophy of life, but also with nationalism and industrialization. Indeed, it appears that 'activism' was an economic strategy originally employed for the benefit of the nation-state in which its citizens could enjoy the largest amount of 'spiritual freedom' through actively encouraged but closely monitored economic competition.

Activism thus had several different meanings in its history—a philosophical orientation to life, an economic strategy to mobilize citizens for national industrialization, a pro-German activity during World War I, and a vigorous political activity.

In its contemporary usage, *activism* generally refers to citizens' political activities ranging from high-cost, high-risk protests and revolutionary movements (McAdam 1986) to everyday practices aimed at protecting the environment (Almanzar, Sullivan-Catlin, and Deane 1998) and to corporatized NGO activism (Spade 2011). Its popularity undoubtedly has something to do with its multiple, ambiguous meanings, which make the word suitable for different purposes.

Over the past thirty years, *activism* has also become less likely to mean radical and revolutionary action and more likely to mean moderate civic action. Many social movements and activism studies support this hypothesis (Meyer and Tarrow 1998; Samson et al. 2005; Spade 2011). A glance at the frequency of the term and its associated words also helps: consider Google Ngram Viewer, which contains 5.2 million scanned books published between 1550 and 2008, with 500 billion words in total and 361 billion in English (Michel et al. 2010). As a hypothesis, suppose that a stronger association of *activism* with *revolution* or *protest* rather than *NGO* or *civil society* implies a more radical connotation, whereas a declining association with terms like *revolution* may indicate a less radical connotation. Now observe in figures 1 and 2 the patterns of usage in Google Ngram Viewer for *activism* in comparison with *revolution* and *protest* and with *NGO* and *civil society* from 1950 to 2008.

Figure 1 shows that the use of *revolution* declined steadily after the 1970s in proportion with the rising frequency of *activism*, with *protest* holding relatively steady. Figure 2 shows a remarkable

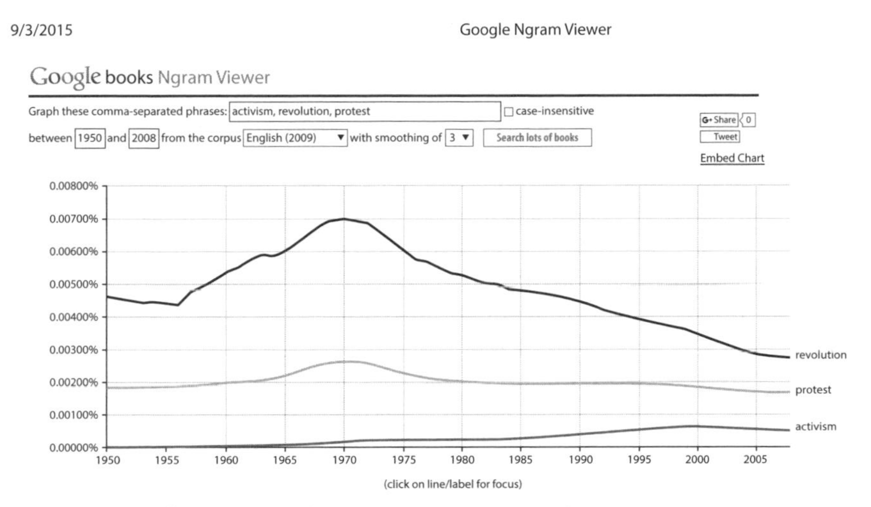

Figure 1. Use of *activism*, *revolution*, and *protest* in Google Ngram Viewer, 1950–2008.

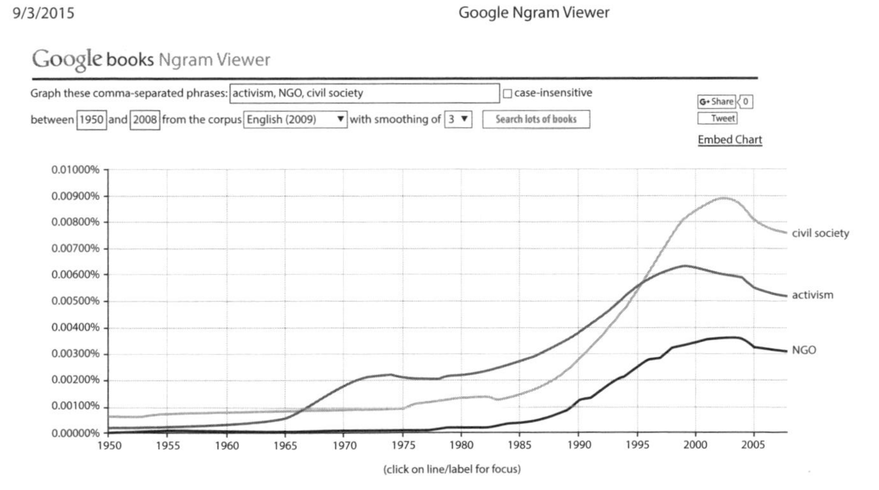

Figure 2. Use of *activism*, *NGO*, and *civil society* in Google Ngram Viewer, 1950–2008.

parallel rise in the use of *activism*, *NGO*, and *civil society*. If NGO and civil society activism tends to be moderate, institutionalized, and even corporatized (Samson et al. 2005; Spade 2011; Dauvergne and LeBaron 2014), rather than radical and revolutionary, then the usage patterns suggest that from 1950 to 2008, especially after the 1990s, *activism* has mellowed to indicate moderate rather than radical forms of action.

An Ambivalent Age

The ambiguity of the increasingly popular keyword appears to serve well the politics and purposes in the current age of ambivalence. That ambivalence is a condition of modernity is already a thesis well developed in the works of classic social theorists from Marx to Weber (Smart 1999), although the post-1989, post–Cold War world entered a period of "new ambivalence" (Beck 1997). This "new ambivalence" rests on the unmooring of traditions and traditional communities, the breakdown of old boundaries of the public and the private, the collapse of faith in human progress, the retreat of grand, emancipatory politics and the rise of life politics, and what Beck calls the "reversal of politics and non-politics" where "the political becomes non-political and the non-political political" (Beck 1992: 186). The causes for this upheavel include, in brief, the shift from industrialization to postindustrialization in global economies, the crisis of the nation-state under the onslaught of globalization, the diffusion of new forms and patterns of communication associated with the development of internet and mobile technologies, and, over and above all these, the disembedding of institutions and the advent of a society of risk (Giddens 1990; Beck 1992). In essence, as Beck puts it, this new ambivalence is the consequence of the "gradual or eruptive collapse of previously applicable basic certainties" for practically all fields of social activity (Beck 1997, 11).

These influences express themselves in political participation complexly. On the one hand, aspirations for political struggle continue to take both radical and nonradical forms. The Zapatista revolt and the protests against the World Trade Organization (WTO) in Seattle in the 1990s, for example, were radical eruptions—as were the more recent Occupy Wall Street and Arab Spring protests. On the other hand, the history of activism and protest since the 1990s remains marked more by moderation than by radicalism in both Western democracies and other countries.

In Western democracies, popular political radicalism declined in the wake of the protest cycle of the 1960s and 1970s. What have appeared instead are "social movement societies," where protest becomes increasingly institutionalized and bureaucratized, and "civic" rather than disruptive. Meyer and Tarrow (1998, 20), editors of the volume *The Social Movement Society*, write, "Although disruption appears to be the most effective political tool of the disadvantaged, the majority of episodes of movement activity we see today disrupt few routines." A study of over four thousand events in the Chicago area from 1970 to 2000 finds that "sixties-style" protest declined, while hybrid events combining public claims making with civic forms of behavior increased (Samson et al. 2005). The most distinctive pattern of the post-1970s landscape of citizen participation is collective *civic* action, not disruptive action (see **participation**).

In China and the former Soviet bloc, large-scale protest activities declined after the "Velvet Revolution" and the Tiananmen student protests in 1989. History was proclaimed to have ended and revolution a relic of the past. In the wake of the Tiananmen protests, even Chinese intellectuals who had supported the Tiananmen movement bid "farewell to revolution," advocating instead reform as a method of political change and a prominent practice since the 1990s (Li and Liu 1995). Deradicalized civic action, such as NGO activism, also became more common than radical protest as revolutionary aspirations gave way to reformist agendas (Yang 2009).

Of course, moderation does not capture all the ambivalent trends of contemporary activism, such as the rise of the outsourcing of grassroots politics (Fisher 2006), the "nonprofitization" of social movements (Spade 2011, 40), and the corporatization of activism (Dauvergne and LeBaron 2014). Some activist organizations outsource their political campaigns to commercial campaign organizations, much as big corporations outsource their jobs and products to overseas factories. Others collaborate with businesses in pursuit of dubious goals, as when environmental organizations partner with oil companies to protect the environment. Deeply troubling to critics, "activist organizations have increasingly come to look, think, and act like corporations" in the last two decades (Dauvergne and LeBaron 2014, 1).

The Internet Brings Hope

At the same time the internet has introduced a range of new practices known as cyberactivism, hacktivism, internet activism, digital activism, and online activism. Symptomatic of the large world-historic transformations sketched above, these forms promise for many retransformation toward a more radical grassroots politics.

Online or cyberactivism (McCaughey and Ayers 2003) dates back to at least the mid-1990s (Jordan and Taylor 2004, 13) and the hacker communities of the 1980s or earlier (Turkle 1984; Levy 1984; Bowcott and Hamilton 1990). Hacktivism seeks radical ways of "beating the system" in particular (Bowcott and Hamilton 1990). And the Zapatista Army of National Liberation (EZLN) used online electronic bulletin board systems (BBS) and newsgroups for protest in the mid-1990s (Wolfson 2014). During the Tiananmen student protests in 1989, Chinese students in North America and Europe used newsgroups on Usenet to mobilize support for protesters in China (Yang 2009, 28–29). By the mid-1990s, cybercultural and activist communities were proclaiming the power of the new technologies with manifestos: Timothy May issued his "Crypto Anarchist Manifesto" in 1992; in 1996, John Perry Barlow published his "Declaration of the Independence of Cyberspace," while the Critical Art Ensemble (1996) published its manual for electronic civic disobedience; and in 1997, dissidents in China launched what they thought of as the first electronic magazine with the bold statement that computer networks had changed the equation between the autocrats and their struggles for freedom and democracy (Yang 2009, 92).

Other terms have followed the rapid proliferation and diffusion of global network technologies, such as *tactical media* (Garcia and Lovink 1997), *radical media* (Downing 2000), *new media activism* (Kahn and Kellner 2004; Lievrouw 2011), *alternative media* (Couldry and Curran 2003; Lievrouw 2011), *hacktivism* (Denning 1999; Jordan and Taylor 2004; Coleman 2013), *networked social movements* (Castells 2012), and *online connective action* (Bennett and Segerberg 2013). The outcomes of online activism discourse, however, remain far from clear. After over two decades of struggle and contestation, the meanings and significance of online activism are more ambivalent than ever, while the push for its politicization is often offset by the pull toward depoliticization. A closer comparison of these terms reveals both a profound fascination and deep anxieties.

The Ambiguity of Online Activism

Online activism is just as ambiguous as conventional activism. Email and web petitions, the hacking of websites, the Indymedia movement, the Occupy Wall Street movement, and the Arab Spring protests have all been called online or digital activism. But a cyberactivist may also be just a regular computer user seeking to get other people onto the "information superhighway" by any number of everyday activities, such as these ten: (1) join something; (2) use the local library (to go online); (3) respect other people's bandwidth; (4) be all you can be—be "out" online; (5) learn a second language; (6) give your knowledge away free; (7) help a journalist; (8) get your mother on email; (9) encourage a kid; and (10) adopt a Newbie (Coyle 1994).

At other times, one person's online activism could be another's revolution and a third person's crime. The military often views hacking as an act of cyberwarfare, cyberterrorism, or cybercrime (Vegh 2003, 81). Online activism is sometimes dismissed on the ground that it is not as effective as "real" activism on the streets; at other times, street protests are characterized as networked digital protests simply because Facebook and Twitter were used to organize the protests offline. The result is the conflation of the more radical types of online activism with more moderate varieties. Far from a naive confusion of the meanings of a word, this conflation represents a history of political struggle over the meanings and practices of online activism.

Although the trend is neither straightforward nor irreversible, the more radical elements of online activism are becoming underplayed, if not dislodged, in activism discourse. Governments seek to criminalize radical online activists, and corporations seek to co-opt them. Thus over time, *hacktivism* has begun to connote illegality rather than its early evocations of countercultural creativity and individual heroism (Jordan 1999; Taylor 1999; Turner 2008). Radical cyberactivist organizations and practices like Indymedia and the Occupy Wall Street movement were subject to policing (Downing 2000; Pickard 2006; Sullivan, Spicer, and Böhm 2011; Gillham, Edwards, and Noakes 2013; Sauter 2014). In addition, online activism appears increasingly channeled into conventional institutional politics: "The digirati needs to learn how to make friends and win influence in Washington," Richard F. O'Donnell (1996) warned in 1996; "Otherwise they would be courting irrelevance." The online movement Moveon.org, for example, has become a member-based nonprofit organization. Like mirror images (see **mirror**), policing and mainstreaming have sped the absorption of online activism into normal institutional politics, undercutting its subversive potential as an extrainstitutional praxis. As in activism, so in online activism: the early radical aspirations of cyberactivism and hacktivism have weakened.

Online Activism in China

In his introduction to the 1976 edition of *Keywords*, Williams notes a limitation to his book: many keywords have developed important meanings in other languages, but, with few exceptions, these meanings go unnoticed in his analysis. Williams stresses the comparative study of how the keyword meanings go through complicated developments. In this spirit, let us examine a cluster of Chinese words related to online activism to reveal comparative meanings with the English language.

Until recent decades, *activism* was uncommon in the mainland Chinese vocabulary. Its literal translation, *xingdong zhuyi* (行动主义), appeared in academic articles in the late 1980s and early 1990s introducing histories of Western and Korean student activism; the term would not enter everyday language until later.[1] The term *activist*, however, has been a keyword in official Chinese discourse since at least the Mao era: an *activist* actively participates in projects and campaigns sponsored by the government. In other words, an activist actively serves, not opposes, the nation-state. A dissident or protester is the opposite of an activist in this original Chinese sense.

Although varieties of online activism (in the Western sense) began to emerge in China in the 1990s, there were no standard concepts to describe them then. The Chinese word for *protest*, *kangyi*, was sometimes used. One of the most influential online forums in the 1990s and early 2000s, the Strong Nation forum, was originally set up as the "Protest Forum," to rally, for example, against the NATO bombing of the Chinese embassy in the former Yugoslavia in 1999.

Since the late 1990s, several terms have migrated from English into Chinese to describe online activism. Translated back into English, one would mean literally "online movements" (*wangluo yundong*). Another is a literal translation of *online activism* (*wangluo xingdong zhuyi*). Perhaps the most interesting Chinese term is *new media events* (*xin meiti shijian*). An extension of the concept of "media events" in the works of Dayan and Katz (1992) and others, *new media events* is the title of an influential Chinese-language collection of essays about online activism (Qiu and Chan 2011) (see **event**). The term has legs in Chinese academic and media discourse, partly because it avoids the charged political connotations of online activism versus online protest and thus suits China's political context.

Government officials soon invented their own term—*internet mass incidents*—for describing, managing, and containing online activism, which is modeled after the term *mass incidents*, an official label for protest activities. *Mass incidents* and *internet mass incidents* carry clear negative meanings as part of a cluster of new official labels coined for talking about and managing online activism.

As in Western democracies, so in Chinese *online activism*: ambivalence abounds. Perhaps the best example of discursive uncertainty is the term *human flesh search*. This ghastly phrase was first the name of a popular online forum that mobilized and crowdsourced users, online and off, to search for solutions to queries of interest to the forum and community members (see **community**, **forum**). The term then came to be applied to a large number of users pooling their collective wisdom to carry out a specific task, such as exposing a corrupt government official by revealing his private information.

Like doxxing, the online practice of "compiling and releasing a dossier of personal information on someone" (Honan 2014), the means of this form of activism is morally ambiguous, ranging from online vigilantism with disturbing violations of privacy to an outlet for healthful expressions of social grievances and discontent.

Some US-China Comparisons

The Chinese government's approach to online activism suggests that online activism appears more radical and subversive in China than in the United States. This is puzzling, since the United States is a democracy and China not, and since the Chinese government censors the internet and the United States does not.[2] I will argue that this follows because online activism in the United States has been more institutionalized, whereas in China it remains largely noninstitutional or extrainstitutional in practice. Differences in political systems as well as political culture help unravel this puzzle.

For example, most cases of online activism in China occur by means of unorganized, individual participation, whereas in the United States, activism that is primarily confined to the internet is more likely to be dismissed as a "lazy" activism, dubbed as slacktivism or clicktivism. Activism in the West without the support of established civil society organizations is more likely to be viewed as ineffective.

Some scholars counter that since clicktivism, such as mass emails sent by advocacy organizations, is only a single tactic in a whole repertoire of action used by advocacy groups, it is unhelpful to understand it in isolation from the rest of the advocacy repertoire (Karpf 2012). Even this valid point, however, tacitly acknowledges that digital-only activism sits on the lowest rung of the citizen participation reputation ladder.

If there is a ladder of engagement in China, then the online type of internet activism would rank much higher there than in the United States. In China, ordinary internet users and dissidents alike see the internet as essential to political expression and use it as such. When they protest online, their voices may resonate enough to challenge government policies and officials' behavior. For this reason, online protests, officially labeled as *internet mass incidents*, are viewed as serious threats to the legitimacy of the party-state and are carefully censored and controlled.

These differences reflect the relative strength of institutionalized social movements and civil society in the United States compared to the same in China. In China, online activism was born into a weak civil society in the late 1990s, when NGOs were beginning to emerge, but with only minimal influence in interest articulation. Street protests faced serious risks, and the state owned and controlled the mass media. Meanwhile, public frustration and anger over social injustices and government corruption were deepening. Under these circumstances, citizens took to the internet to sound their voice: by the early 2000s, a culture of online contention had formed in China, but one in which individual citizens participate in unorganized ways rather than through civil society organizations.

By contrast, in the United States, online activism arose embedded in a long tradition of nonprofit, community, and social movement organizations. Online activism in the United States is only one means by which nonprofit and civil society organizations recruit members, raise funds, conduct publicity campaigns, and build communities (Hick and McNutt 2002; McNutt and Menon 2008; Guo and Saxton 2014). United for Peace and Justice, for example, organized the 400,000-strong protest against the war in Iraq on January 27, 2007, using the web (Earl and Kimport 2011). The news here is not so much the web as the UFPJ, a coalition of over 1,300 groups.

Given this institutional context, spontaneous and unorganized forms of online action—or what would be termed *internet mass incidents* by Chinese officials—are also likely to be viewed with suspicion and moral alarm by the American public.[3] For example, the unorganized but collective efforts in 4Chan and Reddit online communities to search for the Boston bombing suspects after April 15, 2013, a kind of online collective action not unlike the online muckraking directed against corrupt government officials in China, was met with public criticism and cries of vigilantism (see, e.g., Madrigal 2013). It is a most curious fact, then, that Chinese officials and the American public share a similar hermeneutic of suspicion toward unorganized new media events.

Transforming Activism after 2010?

How have the Arab Spring protests that started in 2010 and the global "Occupy" movement in 2011 affected the meaning of *activism*?

Despite disagreements about the political outcomes of these protests, there are clear signs that both movements, in their revolutionary scale and speed of popular mobilization worldwide, have reinfused radicalism and vigorous political activity back into the English sense of *activism*. They adopted radical forms of action, most notably the sustained occupation of central public spaces and demonstrations in cities around the world, from New York to Cairo to Madrid. And despite the refusal to make any programmatic demand in the case of the Occupy Wall Street movement, the movements aspired to systematic and thus revolutionary change. And certainly, since the Arab Spring and the "Occupy" movement, the language of *revolution* has reentered public discourse with a refreshed sense of moral legitimacy. From *Facebook Revolution* and *Twitter Revolution* to *networked revolution* and the *revolution of dignity*, the language of revolution has suffused the language of activism and protest movements (Castells 2012; Gitlin 2013).

The radical means of the Arab Spring protests are clear,[4] given that governments were overthrown and blood was shed. But what about Occupy Wall Street (OWS)? In discussing the radicalism of OWS, analysts have pointed to its dedication to free assembly (Gitlin 2013), to the populist expressions of feelings, passions, and invocations of "the people" (Calhoun 2013), to the ideal of direct democracy (Castells 2012), and to the practice of a new form of "political disobedience" (Harcourt 2013).

Encapsulated in these features of the radical "Occupy" movement is a conundrum of contemporary activism.[5] As Harcourt explains, political disobedience is "a type of political as opposed to civil disobedience that fundamentally rejects the ideological landscape that has dominated our collective imagination, in the United States at least, since before the Cold War" (Harcourt 2013, 46). It "resists the very way in which we are governed":

> It disobeys not only our civil structure of laws and political institutions, but politics writ large. The Occupy movement

> rejects conventional political rationality, discourse, and strategies. It does not lobby Congress. It defies the party system. It refuses to align or identify itself along traditional political lines. It refuses even to formulate a reform agenda or to endorse the platform of any existing political group. Defying convention, it embraces the idea of being "leaderless." It aspires to rhizomic, nonhierarchical governing structures. And it turns its back on conventional political ideologies. Occupy Wall Street is politically disobedient to the core—it even resists attempts to be categorized politically. The Occupy movement, in sum, confounds our traditional understandings and predictable political categories. (Harcourt 2013, 47)

Perhaps nothing can be more radical than an overt rejection of and abstinence from the existing system: the "Occupy" movement may be transforming the contemporary meaning of *activism*.

In order to resist being "categorized politically," the "Occupy" movement has sought to express itself in nonconventional political forms. A new political disobedience turns out to be nonpolitical in the conventional sense. Harcourt's analysis is convincing, but in it we encounter a conundrum facing radical revolutionaries and activists under the current capitalist hegemony. And this is the same conundrum that faces many NGO activists engaged in today's "conventional" political activism: to bring about change, one has to work with the system; NGOs have to play by the rules rather than against them; organizations have to ally with rather than against governments and corporations. The result is often cooptation. Especially since the heightened securitization of activism and dissent after 9/11 (Dauvergne and LeBaron 2014), it is increasingly difficult for social movement organizations to work against the system. Given all this, it is too early to tell whether the global wave of popular protests since 2010 has transformed or radicalized what we mean by *activism*. Its effects will likely remain temporary and ambivalent.

See in this volume: cloud, community, democracy, event, mirror, participation, personalization, sharing

See in Williams: liberation, progressive, radical, reactionary, reform, revolution, violence

Notes

1 From 1991 to 1993, the Chinese academic journal *Youth Studies* (*qingnian yanjiu*) published a series of translated articles on student activism in foreign countries, probably to help Chinese researchers better understand Chinese student activism.
2 Of course, it is necessary to distinguish between internet censorship and surveillance, noting the post-Snowden revelations about the widespread practices of online surveillance in the United States.
3 The rise of hashtag activism, however, may signal the growing significance of individual civic participation on social media. See Clark (2014).
4 But see Bayat (2013, 599) for a forceful argument that the Arab Spring protests were not revolutions, but "refo-lutions," by which he means "revolutionary movements that wished to compel the incumbent regimes to reform themselves."
5 I thank Rosemary Clark and Jonathan Pace for sharing their views about Harcourt's analysis of "political disobedience" in our extended discussions about the "Occupy" movement in the spring of 2014.

References

Almanzar, Nelson A. Pichardo, Heather Sullivan-Catlin, and Glenn Deane. 1998. "Is the Political Personal? Everyday Behaviors as Forms of Environmental Movement Participation." *Mobilization* 3(2): 185–205.
Bayat, Asef. 2013. "The Arab Spring and Its Surprises." *Development and Change* 44(3): 587–601.
Beck, Ulrich. 1992. *Risk Society: Towards a New Modernity*. London: Sage Publications.
Beck, Ulrich. 1997. *The Reinvention of Politics: Rethinking Modernity in the Global Social Order*. Cambridge: Polity Press.
Bennett, Lance W., and Alexandra Segerberg. 2013. *The Logic of Connective Action: Digital Media and the Personalization of Contentious Politics*. Cambridge: Cambridge University Press.
Bowcott, O., and S. Hamilton. 1990. *Beating the System*. London: Bloomsbury.
Calhoun, Craig. 2013. "Occupy Wall Street in Perspective." *British Journal of Sociology* 64: 26–38. doi: 10.1111/1468-4446.12002.
Castells, Manuel. 2012. *Networks of Outrage and Hope: Social Movements in the Internet Age*. Cambridge: Polity Press.
Clark, Rosemary. 2014. "#NotBuyingIt: Hashtag Feminists Expand the Commercial Media Conversation." *Feminist Media Studies* 14(6): 1108–10.
Coleman, Gabriella E. 2013. *Coding Freedom: The Ethics and Aesthetics of Hacking*. Princeton, NJ: Princeton University Press.
Couldry, Nick, and James Curran, eds. 2003. *Contesting Media Power. Alternative Media in a Networked World*. Lanham, MD: Rowman and Littlefield.

Coyle, Karen. 1994. "Karen Coyle's List of 10 Things You Can Do as a Cyber-Activist." Talk given at DEFCON, August, Las Vegas. http://www.kcoyle.net/deftlk.html.

Critical Art Ensemble. 1996. *Electronic Civil Disobedience and Other Unpopular Ideas*. New York: Autonomedia.

Dauvergne, Peter, and Genevieve LeBaron. 2014. *Protest Inc.: The Corporatization of Activism*. Cambridge: Polity Press.

Dayan, Daniel, and Elihu Katz. 1994. *Media Events: The Live Broadcasting of History*. Cambridge, MA: Harvard University Press.

Denning, D. 1999. "Activism, Hacktivism, and Cyberterrorism: The Internet as a Tool for Influencing Foreign Policy." Available at http://www.rand.org/content/dam/rand/pubs/monograph_reports/MR1382/MR1382.ch8.pdf.

Downey, Tom. 2010. "China's Cyberposse." *New York Times*. March 3. Available at http://www.nytimes.com/2010/03/07/magazine/07Human-t.html?pagewanted=all&module=Search&mabReward=relbias%3As%2C%7B%222%22%3A%22RI%3A14%22%7D&_r=0.

Downing, John. 2000. *Radical Media: Rebellious Communication and Social Movements*. London: Sage Publications.

Earl, Jennifer, and Katrina Kimport. 2011. *Digitally Enabled Social Change: Activism in the Internet Age*. Cambridge, MA: MIT Press.

Fisher, Dana. 2006. *Activism, Inc.: How the Outsourcing of Political Campaigns Is Strangling Progressive Politics in America*. Stanford, CA: Stanford University Press.

Garcia, David, and Geert Lovink. 1997. "The ABC of Tactical Media." www.ljudmila.org/nettime/zkp4/74.htm.

Giddens, Anthony. 1990. *The Consequences of Modernity*. Stanford, CA: Stanford University Press.

Gillham, Patrick F., Bob Edwards, and John A. Noakes. 2013. "Strategic Incapacitation and the Policing of Occupy Wall Street Protests in New York City." *Policing and Society: An International Journal of Research and Policy* 23(1). doi: 10.1080/10439463.2012.727607.

Gitlin, Todd. 2013. "Occupy's Predicament: The Moment and the Prospects for the Movement." *British Journal of Sociology* 64: 3–25. doi: 10.1111/1468-4446.12001.

Guo, Chao, and Gregory D. Saxton. 2014. "Tweeting Social Change: How Social Media Are Changing Nonprofit Advocacy." *Nonprofit and Voluntary Sector Quarterly* 43(1): 157–79.

Harcourt, Bernard E. 2013. "Political Disobedience." In W.J.T. Mitchell, Bernard E. Harcourt, and Michael Taussig, *Occupy: Three Inquiries in Disobedience*, 45–92. Chicago: University of Chicago Press.

Hick, Steven, and John G. McNutt, eds. 2002. *Advocacy and Activism on the Internet: Perspectives from Community Organization and Social Policy*. Chicago: Lyceum Books.

Honan, Mat. 2014. "What Is Doxxing?" *Wired*. March 6. http://www.wired.com/2014/03/doxing/.

Hoofd, Ingrid M. 2008. "Complicit Subversions: Cultural New Media Activism and 'High' Theory." *First Monday* 13(10). October 6. http://firstmonday.org/ojs/index.php/fm/article/view/2147/2032.

Jordan, Tim. 1999. *Cyberpower: The Culture and Politics of Cyberspace and the Internet*. London: Routledge.

Jordan, Tim, and Paul Taylor. 2004. *Hactivism and Cyberwars: Rebels with a Cause*. London: Routledge.

Kahn, Richard, and Douglas Kellner. 2004. "New Media Activism: From the 'Battle of Seattle' to Blogging." *New Media & Society* 6(1): 87–95.

Karpf, David. 2012. *The MoveOn Effect: The Unexpected Transformation of American Political Advocacy*. Oxford: Oxford University Press.

Lievrouw, Leah A. 2011. *Alternative and Activist New Media*. London: Polity Press.

Li Zehou and Liu Zaifu. 1995. *Gaobie gemin* [Farewell to revolution]. Hong Kong: Tiandi tushu.

Levy, Steven. 1984. *Hackers: Heroes of the Computer Revolution*. New York: Bantam Doubleday Dell.

Madrigal, Alexis C. 2013. "Hey Reddit, Enough Boston Bombing Vigilantism." *Atlantic*, April 17. http://www.theatlantic.com/technology/archive/2013/04/hey-reddit-enough-boston-bombing-vigilantism/275062/.

McAdam, Doug. 1986. "Recruitment to High-Risk Activism: The Case of Freedom Summer." *American Journal of Sociology* 92(1): 64–90.

McCaughey, Martha, and Michael D. Ayers. 2003. *Cyberactivism: Online Activism in Theory and Practice.* New York: Routledge.

McNutt, John G., and Goutham M. Menon. 2008. "The Rise of Cyberactivism: Implications for the Future of Advocacy in the Human Services." *Families in Society: The Journal of Contemporary Social Services* 89(1): 33–38.

Meyer, David S., and Sidney G. Tarrow. 1998. *The Social Movement Society: Contentious Politics for a New Century*. Lanham, MD: Rowman & Littlefield.

Michel, Jean-Baptiste, Yuan Kui Shen, Aviva Presser Aiden, Adrian Veres, Matthew K. Gray, William Brockman, The Google Books Team, Joseph P. Pickett, Dale Hoiberg, Dan Clancy, Peter Norvig, Jon Orwant, Steven Pinker, Martin A. Nowak, and Erez Lieberman Aiden. 2010. "Quantitative Analysis of Culture Using Millions of Digitized Books." Published online ahead of print, December 16. *Science* 331(6014) (2011): 176–82.

O'Donnell, Richard F. 1996. "Courting Irrelevance: The Digirati Needs to Learn How to Make Friends and Win Influence in Washington." Progress & Freedom Foundation, May 3. https://w2.eff.org/Censorship/Internet_censorship_bills/pff_online_activism.critique.

Pickard, Victor. 2006. "United yet Autonomous: Indymedia and the Struggle to Sustain a Radical Democratic Network." *Media, Culture & Society* 28(3): 315–36.

Qiu, Linchuan, and Joseph Chan, eds. 2011. *Xin Meiti Shijian Yanjiu* [Studies of new media events)] Beijing: Renmin University Press.

Samson, Robert J., Doug McAdam, Heather MacIndoe, and Simón Weffer-Elizondo. 2005. "Civil Society Reconsidered: The Durable Nature and Community Structure of Collective Civic Action." *American Journal of Sociology* 111(3): 673–714.

Sauter, Molly. 2014. *The Coming Swarm: DDOS Actions, Hacktivism, and Civil Disobedience on the Internet*. New York: Bloomsbury Academic.

Smart, Barry. 1999. *Facing Modernity: Ambivalence, Reflexivity, and Morality*. London: Sage Publications.

Spade, Dea. 2011. *Normal Life: Administrative Violence, Critical Trans Politics, and the Limits of Law*. Cambridge, MA: South End Press.

Sullivan, Sian, André Spicer, and Steffen Böhm. 2011. "Becoming Global (Un) Civil Society: Counter-Hegemonic Struggle and the Indymedia Network." *Globalizaions* 8(5): 703–17.
Taylor, Paul. 1999. *Hackers: Crime and the Digital Sublime*. London: Routledge.
Turkle, Sherry. 1984. *The Second Self: Computers and the Human Spirit.* London: Granada.
Turner, Fred. 2008. *From Counterculture to Cyberculture: Stewart Brand, the Whole Earth Network, and the Rise of Digital Utopianism*. Chicago: University of Chicago Press.
Vegh, Sandor. 2003. "Classifying Forms of Online Activism: The Case of Cyber-protests against the World Bank." In *Cyberactivism: Online Activism in Theory and Practice*, edited by Martha McCaughey and Michael D. Ayers, 71–95. New York: Routledge.
Wolfson, Todd. 2014. *Digital Rebellion: The Birth of the Cyber Left*. Chicago: University of Illinois Press.
Yang, Guobin. 2009. *The Power of the Internet in China: Citizen Activism Online*. New York: Columbia University Press.

2

Algorithm

Tarleton Gillespie

In *Keywords*, Raymond Williams highlights how important terms change over time. But for many of the "digital keywords" here, just as important is the simultaneous use of a term by different communities, particularly inside and outside of technical professions, who seem often to share common words but speak different languages. Williams points to this concern too: "When we come to say 'we just don't speak the same language' we mean something more general: that we have different immediate values or different kinds of valuation, or that we are aware, often intangibly, of different formations and distributions of energy and interest" (1976/1983, 11).

In the case of *algorithm*, the technical specialists, the social scientists, and the broader public are using the word in different ways. For software engineers, algorithms are often quite simple things; for the broader public they are seen as something unattainably complex. For social scientists, *algorithm* lures us away from the technical meaning, offering an inscrutable artifact that nevertheless has some elusive and explanatory power (Barocas, Hood, and Ziewitz 2013, 3). We find ourselves more ready to proclaim the impact of algorithms than to say what they are. This is not to say that critique requires a settled, singular meaning, or that technical meanings necessarily trump others. But we should be cognizant of the multiple meanings of *algorithm* as well as the discursive work the term performs.

To chase the etymology of the word is to chase a ghost. It is often said that the term *algorithm* was coined to honor the contributions of ninth-century Persian mathematician Muḥammad ibn Mūsā al-Khwārizmī, noted for having developed the fundamental techniques of algebra. It is probably more accurate to say that it developed from or with the word *algorism*, a formal term for the Hindu-Arabic decimal number system, which was sometimes spelled *algorithm*, and which itself is said to derive from a French bastardization of a Latin bastardization of al-Khwārizmī's name, *Algoritmi*. Either way, it is something beyond irony that *algorithm*, which now drops its exotic flavor into Western discussions of the information society, honors an Arabic mathematician from the high court of Baghdad. The decimal number system he helped popularize also introduced the concept of zero, or *sifr* in Arabic. Perhaps it is fitting that al-Khwārizmī also has a crater on the moon named after him, a kind of astronomic zero. Like his crater and the zero concept he championed, the term *algorithm* will turn out to be important in part because it is vacant, a cypher, a ghostly placeholder upon which computational systems now stand.

Algorithm as a Trick

As we try to pinpoint the values that are enacted, or even embedded, in computational technology, it may in fact not be the algorithms that we need be most concerned about—if what we meant by *algorithm* was restricted to software engineers' use of the term. For the makers of algorithms, the term refers specifically to the logical series of steps for organizing and acting on a body of data to quickly achieve a desired outcome. MacCormick (2012), explaining algorithms to a general audience, calls them "tricks" (5), by which he means "tricks of the trade" more than tricks in the magical sense—or perhaps like magic, but as a magician understands it. An algorithm is a recipe composed in programmable steps; most of the "values" that concern us lie elsewhere in the technical systems and the work that produces them.

For its designers, the algorithm comes only after the generation of a "model." The model is the formalization of a problem and its goal, articulated in computational terms. So the goal of *giving users the most relevant search results for their query* might be modeled as, or approximated into operationalized terms as, *efficiently calculating the combined values of preweighted objects in the index database, in order to improve the percentage likelihood that users click on one of the first five results.*[1] The complex social activity and the values it holds dear are translated into a functional interaction of variables, steps, and indicators. What was a social judgment—"What's relevant?"—gets modeled: posited and measurable relationships, actionable and strategic targets, and threshold indicators of success.

The algorithm, then, is merely the procedure for addressing the task as operationalized: steps for aggregating those assigned values efficiently, or making the matches rapidly, or identifying the strongest relationships according to some operationalized notion of "strong." All is in the service of the model's understanding of the data and what they represent, and the model's goal and how it has been formalized. There may be many algorithms that would reach the same result inside a given model, just as bubble sorts and shell sorts will both alphabetize lists of words successfully. Engineers choose between them based on "technical" values such as speed, system load, perhaps their computational elegance. The embedded values that make a sociological difference are probably more about the problem being solved, the way it has been operationalized, the goal chosen, and the way that goal has been operationalized (Rieder 2012).

Of course, simple alphabetical sorting is a misleading example to use here. The algorithms we're concerned about today are rarely designed to reach a single and certifiable answer, like a correctly alphabetized list. Most common algorithms produce no certifiably "correct" results at all but only turn out results based on many possible pathways. Algorithm designers are not pursuing correctness; they're pursuing some threshold of operator or user satisfaction—understood in the model, perhaps, in terms of percent clicks on the top results; or percentage of correctly identified human faces from digital images.

Contemporary algorithms, especially those involved in some form of machine learning, are also "trained" on a corpus of existing data. These data have been in some way certified, either by the designers or by past user practices: this photo is of a human face, this photo is not; this search result has been selected by many users in response to this query, this one has not. The algorithm is then run on these data so that it may "learn" to pair queries and results found satisfactory in the past, or to distinguish images with faces from images without.

The values and assumptions that go into the selection and preparation of these training data may be of much more importance to our sociological concerns than the algorithm that's learning from them. For example, the training data must be a reasonable approximation of the data that algorithm will operate on in the wild. The most common problem in algorithm design is that the training data turn out not to match the data being operated on in the wild, in some consequential way. Sometimes new phenomena emerge that the training data simply did not include and could not have anticipated; just as often, something important was overlooked as irrelevant, or was scrubbed from the training data in preparation for the development of the algorithm. Imagine a recognition algorithm trained on a corpus of selfies, but the photo archive came from an online service that is used disproportionately by people of particular races. The algorithm designed may later prove less accurate with a more diverse corpus of photos, and may therefore seem to have deeply problematic implications.[2]

Furthermore, improving an algorithm is rarely about redesigning it. Rather, designers "tune" an array of parameters and thresholds, each of which represents a tiny assessment or distinction. In search, this might mean the weight given to a word based on where it appears in a webpage, or assigned when two words appear in proximity, or given to words that are categorically equivalent to the query term. These thresholds can be dialed up or down in the algorithm's calculation of which webpage has a score high enough to warrant ranking it among the results returned to the user.

Finally, these exhaustively trained and finely tuned algorithms are instantiated inside of what we might call an application. For software engineers, the algorithm is the conceptual sequence of steps, which should be expressible in any computer language, or in human or logical language. They are then instantiated in code, running on servers somewhere, attended to by other helper applications (Geiger 2014), and triggered when a query comes in or an image is scanned. These applications may embody values as well, outside of their reliance on a particular algorithm.

To inquire into the implications of algorithms, if we mean what software engineers mean by the term, could only mean something so picky as investigating the political implications of using a bubble sort or a shell sort—and perhaps missing the bigger questions, like why alphabetical in the first place, or why train on this particular data set. Perhaps there are lively insights to be had about the implications of different algorithms in this strict technical sense,[3] but by and large we in fact mean something else when we talk about an algorithm having "social implications."

Algorithm as Synecdoche

While it is important to understand the technical specificity of the term, *algorithm* has now achieved some purchase in the broader public discourse about information technologies, where it is typically used as an abbreviation for everything described above, combined: algorithm, model, target goal, data, training data, application, hardware. As Goffey puts it, "Algorithms act, but they do so as part of an ill-defined network of actions upon actions" (2008, 19). It is this ill-defined network to which our more common use of the term refers. And this technical assemblage stands in for, and often obscures, the people involved at every point: people debating the models, cleaning the training data, designing the algorithms, tuning the parameters, deciding on which algorithms to depend on in which context. "These algorithmic systems are not standalone little boxes, but massive, networked ones with hundreds of hands reaching into them, tweaking and tuning, swapping out parts and experimenting with new arrangements. . . . We need to examine the logic that guides the hands" (Seaver 2013). Perhaps *algorithm* is coming to serve as the name for a particular kind of sociotechnical ensemble, one of a family of systems for knowledge production or decision making: in this one, people, representations, and information are rendered as data, are put into systematic/mathematical relationships with each other, and then are assigned value based on calculated assessments about them.

But what is gained and lost by using *algorithm* this way? Calling the complex sociotechnical assemblage an *algorithm* avoids the need for the kind of expertise that could parse and understand the different elements; a reporter may not need to know the relationships between model, training data, thresholds, and application in order to call into question the impact of that "algorithm" in a specific instance. It also acknowledges that, when designed well, an algorithm is meant to function seamlessly as a tool; perhaps it can, in practice, be understood as a singular entity. Even algorithm designers, in their own discourse, shift between the more precise meaning and this broader use.

On the other hand, this conflation risks obscuring the ways in which political values may slip in elsewhere than at what designers call the *algorithm*. This helps account for the way many algorithm designers seem initially surprised by the interest of sociologists in what they do—because they may not see the values in their "algorithms" (precisely understood) that we see in their algorithms (broadly understood), because questions of value are very much bracketed in the early decisions about how to operationalize a social activity into a model, and lost in the minuscule, mathematical moments of assigning scores and tuning thresholds.

In our own scholarship, this kind of synecdoche is perhaps unavoidable. Unexamined, it reifies the very processes that constitute it. It is too easy to treat it as a singular artifact, when in the cases we're most interested in, it's rarely one tool, but many tools functioning together, sometimes different tools for different users, so complex that in some cases even their designers can no longer comprehend them.[4] It also tends to erase the people involved, downplay their role, and distance them from accountability. In the end, whether this synecdoche is acceptable depends on our intellectual aims. Calling all these social and technical elements the *algorithm* may give us a handle with which to grip what we want to closely interrogate; at the same time it can produce a "mystified abstraction" (Striphas 2012) that, for other research questions, it might be better to demystify.

Algorithm as Talisman

The information industries often invoke the term *algorithm* to the public as well. To call a service or process an *algorithm* is to lend it a set of associations: mathematical, logical, impartial, consistent. Algorithms seem to have a "disposition towards objectivity" (Hillis, Petit, and Jarrett 2013, 37); this objectivity is regularly performed as a feature of algorithmic systems (Gillespie 2014). Conclusions described as having been generated by an algorithm wear a powerful legitimacy, much the way statistical data bolster scientific claims. It is a different kind of legitimacy from one that rests on the subjective expertise of an editor or a consultant, though it is important not to assume that it trumps such claims in all cases. A market prediction that is "algorithmic" is different from a prediction that comes from expert brokers highly respected for their expertise and acumen; a claim about an emergent social norm in a community generated by an algorithm is different from one generated ethnographically. Each makes its own play for legitimacy, and implies its own framework for what legitimacy is (quantification or interpretation, mechanical distance or human closeness) (see community). But in the context of nearly a century of celebration of the statistical production of knowledge and long-standing trust in automated calculation over human judgment, the algorithmic does enjoy a particular cultural authority.

More than that, the term offers the corporate owner a powerful talisman to ward off criticism, when companies must justify themselves and their services to their audience, explain away errors and unwanted outcomes, and justify and defend the increasingly significant roles they play in public life (Gillespie 2012a). When critics say, "Facebook's algorithm," they often mean Facebook and the choices it makes, some of which are made in code. But information services can point to "the algorithm" as having been responsible for particular results or conclusions, as a way to distance those results from the providers (Morozov 2014, 142). The term generates an entity that is somehow separate, like the assembly line inside the factory, that can be praised as efficient or blamed for mistakes.

The term *algorithm* is also quite often used as a stand-in for its designer or corporate owner. This may be another way of making the earlier point, that the singular term stands for a complex sociotechnical assemblage: *Facebook's algorithm* really means *Facebook*, and *Facebook* really means the people, things, priorities, infrastructures, aims, and discourses that animate the site. But it may also be a political economic conflation: this is Facebook acting through its algorithm, intervening in an algorithmic way, building a business precisely on its ability to construct complex models of social/expressive activity, train on an immense corpus of data, tune countless parameters, and reach formalized goals extremely efficiently. Facebook as a company often behaves algorithmically.

Maybe saying "Facebook's algorithm" and really meaning the choices made by Facebook the company is a way to assign accountability (Diakopoulos 2013; Ziewitz 2011). It makes the algorithm theirs in a powerful way, reducing the distance some providers put between "them" (their aims, their business model, their footprint, their responsibility) and "the algorithm" (as somehow separate from all that). On the other hand, conflating the algorithmic mechanism and the corporate owner may obscure the ways these two entities are not always aligned. It is crucial that we distinguish between things done by the algorithmic system and things done in other ways, such as the deletion of obscene images from a content platform, which is sometimes performed algorithmically and sometimes manually (Gillespie 2012b). It is crucial to note slippage between a provider's financial or political aims and the way the algorithmic system actually functions. And conflating algorithmic mechanism and corporate owner misses how some algorithmic approaches are common to multiple stakeholders, circulate among practitioners in specific ways, and embody a tactic that exceeds any one implementation.

Algorithm as Committed to Procedure

In recent scholarship, *algorithm* increasingly appears not as a noun but as an adjective. To talk about "algorithmic identity" (Cheney-Lippold 2011), "algorithmic regulation" (O'Reilly 2013), "algorithmic power" (Bucher 2012), "algorithmic ideology" (Mager 2012), "algorithmic culture" (Striphas 2010), or the "algorithmic turn" (Uricchio 2011) is to highlight a social phenomenon that is driven by and committed to algorithmic systems—which include not just algorithms themselves, but also the computational networks in which they function, the people who design and operate them, the data and users on which they act, and the institutions that provide these services.

What we are really concerned with when we invoke the "algorithmic" here is not algorithms per se, but the insertion of procedure into human knowledge and social experience. What makes something algorithmic is that it is produced by or related to an information system committed (both functionally and ideologically) to the computational generation of knowledge or decisions. This requires the formalization of social facts into measurable data, and the "clarification" (Cheney-Lippold 2011) of social phenomena into computational models that operationalize both problem and solution. These often stand in as proxies for human judgment or action, meant to simulate it as nearly as possible. But the "algorithmic" intervenes in terms of step-by-step procedures that one (computer or human) can enact on this formalized information such that it can be computed. This process is automated so that it can happen instantly, repetitively, and across many contexts, away from the guiding hand of its implementers.

This is not the same as suggesting that knowledge is produced exclusively by a machine abstracted from human agency or intervention. Information systems are always swarming with people; we just can't always see them (Downey 2014; Kushner 2013). And an assembly line might be just as "algorithmic" in this sense of the word, or at least the parallels are important to consider. What is central is the commitment to procedure, and the way procedure distances its human operators from both the point of contact with others and the mantle of responsibility for the intervention they make. It is a principled commitment to the "if/then" logic of computation.

Yet what does *algorithmic* refer to, exactly? To put it another way, what is it that is *not* algorithmic? What kind of "regulation" is being condemned as insufficient when Tim O'Reilly calls for "algorithmic regulation"? It would be all too easy to invoke the algorithmic as simply the opposite of what is done subjectively or by hand, or of what can be accomplished only with persistent human oversight, or of what is beholden to and limited by context. To do so would draw too stark a contrast between the algorithm and something either irretrievably subjective (if we are glorifying the algorithmic) or warmly human (if we're condemning it). If "algorithmic" market predictions or search results are produced by a complex assemblage of people, machines, and procedures, what makes their particular arrangement feel different from other ways of generating information, which are also produced by a complex assemblage of people, machines, and procedures? It is imperative that we look more closely at those practices that precede or stand in contrast to those we posit as algorithmic, and recognize how they too combine the procedural and the subjective, the machinic and the human, the measured and the ineffable. And it is crucial that we continue to examine algorithmic systems ethnographically, to explore how the systemic and the ad hoc coexist and are managed within them.

To highlight their automaticity and mathematical quality, then, is not to contrast algorithms to human judgment. It is to recognize them as part of mechanisms that introduce and privilege quantification, proceduralization, and automation in human endeavors. Our concern for the politics of algorithms is an extension of worries about Taylorism and the automation of industrial labor; about actuarial accounting, the census, and the quantification of knowledge about people and populations; and about management theory and the dominion of bureaucracy. At the same time, we sometimes wish for more "algorithmic" interventions when the ones we face are discriminatory, nepotistic, and fraught with error; sometimes procedure is truly democratic.

We rarely get to watch algorithms work; but picture watching complex traffic patterns from a high vantage point: it is clear that this "algorithmic" system privileges the imposition of procedure, and—to even participate in such a complex social interaction—users must in many ways accept it as a kind of provisional tyranny. The elements can be known only in operational terms; every possible interaction within the system must be anticipated; and stakeholders often point to the system-ness of the system to explain success and explain away failure. The system struggles with the tension between the operationalized aims and the way humanity inevitably undermines, alters, or exceeds those aims. The system is designed and overseen by powerful actors, though they appear only at specific moments of crisis. And it's not clear how to organize such complex behavior in any other way and still have it be functional and fair. Commitment to the system and the complex scale at which it is expected to function makes us beholden to the algorithmic procedures that must manage it. From this vantage point, algorithms are merely the latest instantiation of the modern tension between ad hoc human sociality and procedural systemization—but one that is now powerfully installed as the beating heart of the network technologies we surround ourselves with and increasingly depend upon.

See in this volume: community, culture, digital, information, personalization, prototype

See in Williams: bureaucracy, determine, expert, hegemony, industry, institution, jargon, management, mechanical, pragmatic, standards, technology

Notes

1 This parallels Kowalski's well-known definition of an algorithm as "logic + control": "An algorithm can be regarded as consisting of a logic component, which specifies the knowledge to be used in solving problems, and a control component, which determines the problem-solving strategies by means of which that knowledge is used. The logic component determines the meaning of the algorithm whereas the control component only affects its efficiency" (Kowalski 1979, 424). I prefer to use "model" because I want to reserve "logic" for the underlying premise of the entire algorithmic system and its deployment.
2 This may help explain Google's racially charged image labeling blunder in 2015. See Dougherty 2015.
3 See Kockelman 2013 for a dense but superb example.
4 See Christian 2012.

References

Barocas, Solon, Sophie Hood, and Malte Ziewitz. 2013. "Governing Algorithms: A Provocation Piece." http://governingalgorithms.org/resources/provocation-piece/.
Bucher, Taina. 2012. "Want to Be on the Top? Algorithmic Power and the Threat of Invisibility on Facebook." *New Media & Society* 14(7): 1164–80.
Cheney-Lippold, John. 2011. "A New Algorithmic Identity: Soft Biopolitics and the Modulation of Control." *Theory, Culture & Society* 28(6): 164–81.
Christian, Brian. 2012. "The A/B Test: Inside the Technology That's Changing the Rules of Business." *Wired*, April 25. http://www.wired.com/2012/04/ff_abtesting/.
Diakopoulos, Nicholas. 2013. "Algorithmic Accountability Reporting: On the Investigation of Black Boxes." A Tow/Knight Brief. Tow Center for Digital Journalism, Columbia Journalism School. http://towcenter.org/algorithmic-accountability-2/.
Dougherty, Conor. 2015. "Google Photos Mistakenly Labels Black People 'Gorillas'" *Bits Blog, New York Times*, July 1. http://bits.blogs.nytimes.com/2015/07/01/google-photos-mistakenly-labels-black-people-gorillas/.
Downey, Gregory J. 2014. "Making Media Work: Time, Space, Identity, and Labor in the Analysis of Information and Communication Infrastructures."

In *Media Technologies: Essays on Communication, Materiality, and Society*, edited by Tarleton Gillespie, Pablo J. Boczkowski, and Kirsten A. Foot, 141–66. Cambridge, MA: MIT Press.

Geiger, R. Stuart. 2014. "Bots, Bespoke, Code and the Materiality of Software Platforms." *Information, Communication & Society* 17(3): 342–56.

Gillespie, Tarleton. 2012a. "Can an Algorithm Be Wrong?" *Limn* 1(2). http://escholarship.org/uc/item/0jk9k4hj.

———. 2012b. "The Dirty Job of Keeping Facebook Clean." *Culture Digitally*, February 22. http://culturedigitally.org/2012/02/the-dirty-job-of-keeping-facebook-clean/.

———. 2014. "The Relevance of Algorithms." In *Media Technologies: Essays on Communication, Materiality, and Society*, edited by Tarleton Gillespie, Pablo J. Boczkowski, and Kirsten A. Foot, 167–93. Cambridge, MA: MIT Press.

Goffey, Andrew. 2008. "Algorithm." In *Software Studies: A Lexicon*, edited by Matthew Fuller. Cambridge, MA: MIT Press.

Hillis, Ken, Michael Petit, and Kylie Jarrett. 2013. *Google and the Culture of Search*. New York: Routledge.

Kockelman, Paul. 2013. "The Anthropology of an Equation. Sieves, Spam Filters, Agentive Algorithms, and Ontologies of Transformation." *HAU: Journal of Ethnographic Theory* 3(3): 33–61.

Kowalski, Robert. 1979. "Algorithm = Logic + Control." *Communications of the ACM* 22(7): 424–36.

Kushner, Scott. 2013. "The Freelance Translation Machine: Algorithmic Culture and the Invisible Industry." *New Media & Society* 15(8): 1241–58.

MacCormick, John. 2012. *Nine Algorithms That Changed the Future*. Princeton, NJ: Princeton University Press.

Mager, Astrid. 2012. "Algorithmic Ideology: How Capitalist Society Shapes Search Engines." *Information, Communication & Society* 15(5): 769–87.

Morozov, Evgeny. 2014. *To Save Everything, Click Here: The Folly of Technological Solutionism*. New York: PublicAffairs.

O'Reilly, Tim. 2013. "Open Data and Algorithmic Regulation." In *Beyond Transparency: Open Data and the Future of Civic Innovation*, edited by Lauren Goldstein and Lauren Dyson. San Francisco, CA: Code for America Press. http://beyondtransparency.org/chapters/part-5/open-data-and-algorithmic-regulation/.

Rieder, Bernhard. 2012. "What Is in PageRank? A Historical and Conceptual Investigation of a Recursive Status Index." *Computational Culture* 2. http://computationalculture.net/article/what_is_in_pagerank.

Seaver, Nick. 2013. "Knowing Algorithms." Paper presented at Media in Transition 8, Cambridge, MA. http://nickseaver.net/papers/seaverMiT8.pdf.

Striphas, Ted. 2010. "How to Have Culture in an Algorithmic Age." *The Late Age of Print*, June 14. http://www.thelateageofprint.org/2010/06/14/how-to-have-culture-in-an-algorithmic-age/.

———. 2012. "What Is an Algorithm?" *Culture Digitally*, February 1. http://culturedigitally.org/2012/02/what-is-an-algorithm/.

Uricchio, William. 2011. "The Algorithmic Turn: Photosynth, Augmented Reality and the Changing Implications of the Image." *Visual Studies* 26(1): 25–35.

Williams, Raymond. 1976/1983. *Keywords: A Vocabulary of Culture and Society*. 2nd ed. Oxford: Oxford University Press.
Ziewitz, Malte. 2011. "How to Think about an Algorithm: Notes from a Not Quite Random Walk." Discussion paper for Symposium on Knowledge Machines between Freedom and Control, September 29. http://ziewitz.org/papers/ziewitz_algorithm.pdf.

3

Analog

Jonathan Sterne

Sometime in the 1980s, the terms *analog* and *analogue* began to wildly proliferate, a trend that continued into the 1990s. *Analog* is a shortened version of the word *analogue,* consistent with the American trend of shortening English words (and the proliferation of American English on the internet), a practice I continue in this entry by treating the two words as one. It appeared in technical discussions, but also more broadly in cultural journalism, in humanistic writing, and in everyday talk. We would expect as much with words like *digital* or *computer*, given the expansion of computing in everyday life, and the flood of personal computers to hit the market in that decade. But the growth in references to *analog* and *analogue* in the 1990s is telling as well.[1]

It is also the moment that *analog* comes to fully take on its most pervasive contemporary meaning. As Derek Robinson writes in his keyword entry on the term:

> The term "analog" has come to mean smoothly varying, of a piece with the apparent seamless and inviolable veracity of space and time; like space and time admitting infinite subdivision, and by association with them connoting something authentic and natural, against the artificial, arbitrarily truncated precision of the digital (e.g., vinyl records vs. CDs). This twist in the traditional meaning of "analog" is a linguistic relic of a short-lived and now little-remembered blip in the history of technology. (Robinson 2008, 21)

Robinson goes on to give a history of analog computing. But in this entry, I will argue that the proliferation of *analog*'s meaning as "not-digital" or "separate from computers" emerges more from a set of reactions to digital technology than from the engineering field itself. Put another way, an expanded notion of *the analog* as a condition, which now approaches common sense in a whole range of fields—engineering, computer science, media studies, journalism, music fandom, various media arts and humanities—became a useful rhetorical tool for both promotional and critical discussions of digital technology.

The most recent linguistic innovations around the idea of the analog—as the point of contact with the digital and that which lies entirely outside of it—has led to a largely unexamined conceptual expansion of the analog domain in journalism and scholarship alike. There are at least two major problems with this definition. First, *analog* denotes a specific technical process, where one quality is used to represent another. A violin is not an analog technology, but a synthesizer is because of the defined relationships on which its system is based, such as control voltage and oscillator pitch. Second, the entire world outside of digital processing is not analog, because analog represents a particular technocultural relationship to nature. Nature may well be conceived as *having* analogs within it, but it cannot *be* analog.

The idea of *analog* as *everything not-digital* is in fact *newer* than the idea of *digital*. And as I argue below, expanding the idea of *analog* to cover everything that is *not-digital* comes with a cost, because it effectively diminishes the variety of the world as it elevates conceptions of the digital. It mixes very well with all sorts of digital boosterism because it rhetorically figures the primary point of comparison—whether historical, ontological, aesthetic, institutional, or in some other dimension—as between digital technologies and *everything else in the world*. I cannot imagine a more hyperbolic way of figuring digital technologies. The language of *analog as the not-digital world* is also taken up for critique as well as celebration, but it has a similar figurative effect: inflating "the analog" to "the world" limits the options we have for describing natural, cultural, and technological history to one kind of periodization (analog/digital, or maybe preanalog/analog/digital), when in fact there are many different ways to narrate history, and each comes with a different set of purposes. Making "the digital" a historical villain delivers no greater analytical payload than painting it as a hero.[2] This is not to say we cannot generalize about digital technologies, operations, or even culture. Only that if it is counterpoised to an infinitely expanded notion of "the analog," we lose the analytical power of both terms.

To understand the historical meaning of *analog*'s proliferation, we need to get a sense of both the broader meaning of the term and the specific historical meanings that it took on during the 1970s and 1980s. The *Oxford English Dictionary* etymology has the word entering English from the French *analogue*, meaning "a thing that has characteristics in common with another thing." As evidenced by the web of cross-references in the *OED*, the word clearly belongs to a family: *analogous*, *analogon*, *analogate*, all of which descend from Greek and Latin terms for analogy, which later takes on a sense of proportion as well (*OED*, s.v.).

In the *OED*'s account of *analogue*, there appear to be two distinct historical threads that occasionally meet and imitate one another: a natural science thread from chemistry and biology that renders it as a noun, and a technology thread that descends from computing but quickly exceeds it, which is more likely to render it in adjectival or adverbial form. The accompanying table presents some representative definitions from the *OED*'s *analogue* entry. The *OED*'s entries are often somewhat late compared to common usage, but the conservative dates are at least schematically useful. The left column implies morphological relation or structural homology. Social classes in different countries can be analogues of one another; individuals can be analogues of one another; words or phrases can be analogues of one another. Even the soy-based meat and cheese products of the 1966 entry imply a structural replacement in the diet of one biochemical form with another. It is not an accident that across the space of a century, the interface between media ideas and food chemistry ideas moves from technological reproduction and preservation to synthesis and replacement. Where in the nineteenth century ideas about the preservative power of sound recording borrowed their language from canning and embalming (Sterne 2003, 292–301), in the twentieth century, ideas of artificial sound synthesis limn sound creation with food creation. In both fields, a processed world emerges; and the shared cultural histories of signal processing and food processing have yet to be written.

Representative Definitions from the *OED*'s *Analogue* Entry

1808: an extant species corresponding to fossil form 1817: a part of an organism similar in form or function to another part 1835: an animal group having similarities to another unrelated group 1837: a chemical compound with a molecular structure similar to another 1837: a thing or person analogous to another 1966: a synthetic food product manufactured to represent something in nature	1941: a computer that operates with continuously variable qualities that are analogues of qualities being computed 1947: making use of analog computers or signals (media) 1950: analog-to-digital conversion 1959: recording (but only within engineering contexts) 1969: electronic device 1972: timepiece 1976: musical instrument 1979: audio recording (in discussions of music) 1987: the traditional form of something that has computer mediated-counterpart 1993: old-fashioned

The earliest meanings in the right column begin from the same supposition as in biology and chemistry, but jump into the fields of engineering and computation. Analog computing uses variable qualities (of electricity, of water, etc.) to represent the qualities that are being computed. Designers of early machines chose a physical apparatus "whose operations were *analogous* to" the calculations it was meant to perform (Goldstine 1993, 39). In this way, the "analog" is a representation of a thing in the world. As Paul Edwards has noted, the output of analog computers was often "exactly the sort of signals needed to control other machines (e.g., electrical voltages or the rotation of gears)" (Edwards 1996, 67). It is also worth notice that in this early period, analog computers were at least as often referred to as "electronic" computers to distinguish them from human—often women—computers whose job it was to compute (Light 1999). Derek Robinson emphasizes simulation rather than interconnection: from the 1930s on, analog computers "were used by scientists and engineers to create and explore simulation models, hence their name: a model is something standing in analogical relationship to the thing being modeled" (2008, 21).

As Wendy Chun has shown (2011, 104–31), computers are conceived through and built on analogy, in both their hardware and their software. As the discourses of cybernetics developed analogical ways of describing animals and machines, engineers took up this language to describe and imagine the computing devices they were building. Of course, digital technologies have all sorts of models within them, from the skeuomorphs in software interfaces to signal processing math that is meant to imitate older analog devices like synthesizers. But the connections between digital technologies and all the technologies that came before them run at least as deep as the differences. For instance, like some analog computers, most digital computers use voltages to represent numbers (0 or 1): they measure when the voltage passes a threshold, usually 3.3 or 5 volts (though from a standpoint of theoretical computer science, a computer does not have to be a machine that uses voltages to calculate). Modern computers like PCs and laptops operate within the tightest voltage parameters possible, in part because their designers aim to make the variable voltages coursing through them conform as much as possible to the abstractions of binary code. But they also conform to an older standard: Bob Moog's analog synthesizers *also* operated within a range of 0–5 volts (Pinch and Trocco 2002). Similarly, the regulated spinning platter of the hard drive descends from flat disc records and sewing machines, and like those devices, the hard drive needs a mechanism to maintain a consistent spinning speed, and a head mounted on an arm (chew on those metaphors for a moment) to read the data on the spinning disc, somewhat like an old tape deck, gramophone, or optical soundtrack on a strip of celluloid film. It is thus possible to understand a hard drive, and the computer around it, as a mechanism as much as we would understand it as somehow primarily digital (Kirschenbaum 2008). We could say the same for one of the two primary interfaces for computers for much of their history: much has been made of the screen as an interface, but the keyboard is *also* a skeuomorph, taken from an older technology, and a body-wrecking skeuomorph at that (Jain 2006). In these ways, and countless others, so-called digital media are more similar to than different from the devices that came before them.

Starting with the *OED* entries from the 1950s on, *analog* begins to signal something else: that which is *not-digital*, a category initially defined by its point of contact with digital computing that eventually comes to be defined in terms of its noncontact with digital computing. These senses of the term shape how "the analog" comes to be thought. The concept of analog-to-digital conversion is agnostic about how the analog signal is encoded before it reaches the converter.[3] By the 1970s, *analog* is no longer about points of contact with digital technologies, but about contrasts from digital technologies. An analog timepiece is simply not a digital watch—it could be any kind of watch or sundial. An analog audio recording could be made with cylinder, tape, or vinyl; it is simply not digital. The last two entries reveal the extent of this tendency to generalize. "Designating the traditional form of something that has a digital or computer mediated counterpart" generalizes the "not-digital" definition to cover a host of practices that once had nothing to do with digital technologies but now have everything to do with digital processing: retouching photographs, mapmaking, playing games, and writing down notes. In other words, a category as large as "the analog" may group together processes and practices that have as little to do with one another as each does with its supposed relative in the digital domain.[4]

The last definition is called "colloquial" and refers to people "unaware of or unaffected by computer technology or digital communications; outdated, old-fashioned": using tape measures instead of laser measures; traditional grammar and spelling; "technophobes" who don't adopt the latest digital technologies.

These last definitions are the closest to the most common usages of the term in media studies, but they also have a particular lineage. For the 1987 definition, the *OED* cites Stewart Brand's 1987 *The Media Lab: Inventing the Future at MIT.* Brand wrote the book after spending a year at the Media Lab in 1985–86 at Nicholas Negroponte's invitation (Negroponte no doubt hoped that Brand would write the book).[5] Brand's uses of *analog* in the book span several senses. He uses the "not digital" definition early on:

> Telephones, radio, TV and recorded music began their lives as analog media—every note the listener heard was a smooth direct transform of the music in the studio—but each of them is now, gradually, sometimes wrenchingly, in the process of becoming digitized, which means becoming computerized.

> You can see the difference in the different surfaces of long-playing records and compact disks: the records' grooves are wavy lines; the far tinier tracks of CDs are nothing but a sequence of distinct pits. Analog is continuous, digital is discrete. (Brand 1987, 18)

Brand is in fact wrong about the continuous/discrete comparison—his example works with vinyl records or optical sound-on-film but not with sirens, magnetic tape, or player pianos. But his larger usage is common for the time and is repeated later in the book in his discussion of ISDN network lines (versus "older analog equipment" from the telephone company). He quotes Richard Bolt using the even older noun version of analog as substitute: "The [computer] screen is the analog of the room which you and I now share" (144).

The *OED* latches onto the last usage I can find in the book (not counting the index): he quotes Media Lab member Richard Schreiber saying, "It became obvious that digital retouching could be made absolutely undetectable—as opposed to analog retouching (dodging, airbrushing, etc.), which you can almost always see if you look very carefully. If you have a picture represented by a discrete set of numbers, you may not be able to tell that that was not a natural image" (221). In the context of the book, this use of *analog* is not so far from the others, though it does hint at a semantic shift. Both sound recording and mapmaking are technological, and both can be done in ways other than digital. But the nature of their "not-digital" character is quite different. They are not *analog* in the same way. This is an important distinction for us, but of course it was not an important distinction for the engineers at the Media Lab, or for Brand. They were interested in the point of contact between things that lived outside computers and things that lived inside them. *The Media Lab* is about digitization (or, rather, the possible future of digitization), and so all concerns pass through that filter. As a concept, *the analog* expands and blurs in order to give definition to the digital. And it is clear from reading Brand that these various uses of *analog* were already in wide circulation in the engineering and computer science fields of the 1980s. As a term, *analog* circulates freely in the Media Lab depicted in *The Media Lab*. In picking up his definition, the *OED* is late to the party.

If we take the late 1980s usage on its own terms, it is hard not to hear echoes of Walter Benjamin—*analog* is that which withers in the digital age; or even Ferdinand de Saussure—digital is digital because it is not analog. Because all of the *OED*'s entries for its last two definitions of *analog* come from various forms of digital boosterism and new technology journalism—Brand, *Wired Magazine*, *Lifehacker*, newspaper technology columns—these formulations makes sense. But as a category and as a kind of intellectual shadow, *the analog* has expanded far beyond digital boosterism and journalism. Just as *the digital* becomes imaginable as a cultural condition, so too does *the analog*, hence the new noun construction (as opposed to *the analog of* something). If *analog* refers both to things that come into contact with digital technology—probably to be transduced by it—*and* to things outside the domain of digital technology that *do not* come into contact with it, the term expands to cover the whole of reality. This is a problem inasmuch as the word conflates specific technological condition or operation with reality itself. Ted Friedman, quoting critics of compact discs, summarizes the logic this way: "the real world is analog. . . . Digital, by offering the fantasy of precision, reifies the real world. This complaint can be extended to a more global critique of computer culture: the binary logic of computing attempts to fit everything into boxes of zeros and ones, true and false" (Friedman 2005, 43).

We can find this in cultural theories of technology as well. A year before Brand's book came out, and likely influenced by some of the same engineering and computer science thinking, Friedrich Kittler compared the operations of analog sound recording to reality itself. Contrasting Edison's cylinder phonograph to sheet music, Kittler wrote, "Transposition doesn't equal time-axis manipulation. If phonographic playback speed differs from its recording speed, there is a shift not only in clear sound but in entire noise spectra. What is manipulated is the real rather than the symbolic. Long-term acoustic events such as meter and word length are affected as well" (Kittler 1999, 35). At first look, it appears that Kittler is using Lacanian terminology, distinguishing between the symbolic order and the real. For Lacan, the symbolic order is the space of language, representation, meaning, and subject formation, whereas the real is that which resists or exceeds representation (Lacan 1998). Kittler's point here is thus a posthumanist one: sound recording operates on a plane outside of the human subject or interpretation.

And yet there is also a literalism to his interpretation of the machine. Two pages later, Kittler suggests that media directly rely on the laws of physics and physiology (37). Although Kittler does not use the word *analog* anywhere in this discussion, his approach to analog technology appears to follow the logic described by Derek Robinson, as a regime of continuously varying technologies that more accurately access or at least limn reality: in Kittler's world, the cylinder phonograph conforms to the laws of physics. This is a very different "real" from Lacan's, and Kittler's elision of the two accomplishes precisely the intellectual synthesis that allows media theorists to treat analog technologies as closer to nature.

A more explicit philosophical argument for this position can be found in Brian Massumi's *Parables for the Virtual*. He writes that

> the analog is *process*, self-referenced to its own variations. It resembles nothing outside itself. . . . Sensation, always on arrival a transformative feeling of the outside, a feeling of thought, is the being of the analog. It is matter in analog mode. This is the analog in a sense close to the technical meaning, as a continuously variable impulse or momentum that can cross from one qualitatively different medium into another. Like electricity into sound waves. Or heat into pain. Or light waves into vision. Or vision into imagination. Or noise in the ear into music in the heart. Or *the outside coming in*. (Massumi 2002, 135, emphasis in original)

He contrasts the analog as a general mode of being with the digital, which is a highly restricted mode, "a numerically based form of codification (zeroes and ones). As such, it is a close cousin to quantification. Digitization is a numeric way of arraying alternative states so that they can be sequenced into a set of alternative routines. Step after ploddingly programmed step. Machinic habit" (137).

In both Kittler and Massumi we find an odd historical proposition—that analog machines are somehow closer both to the way the human senses work, *and* to the operations of reality itself, than the technologies that preceded or succeeded them.

Viewed with a bit of historiographic distance, this is at once an unsurprising and a fascinating claim. It is unsurprising because the human sciences' most common figurations of reality, the senses, and interfaces all emerged concurrently with the media these terms are used to describe. Conversely, the emergence of technical media in the nineteenth century provided a platform for new descriptions of reality, the senses, and interfaces, many of which are still in use today. As Kittler himself points out, sound recording and cinema emerge alongside modern physics and physiology (see also Hankins and Silverman 1995; Canales 2009). The claim is fascinating because it proposes a truly radical periodization. The claim that analog media are closer to nature proposes an approximately hundred-year period in human history—roughly from the last quarter of the nineteenth century to the last quarter of the twentieth—when the senses and the world were somehow in more harmonious alignment with the workings of media than at any time before or since. The premise behind this is that analog technologies were both preceded and succeeded by technologies of writing—writing and scores in the nineteenth century, and computer code in the twentieth century. That periodization is the philosophical kernel of analog nostalgia. When critics use some permutation of *analog* to apply a hermeneutic of suspicion to *the digital*, they are making an argument about roughly one hundred golden years in human history.

This reading of the analog is, of course, retrospective. In its time, technologies that we now describe as analog (usually after the fact) were more likely to be understood as jarring or artificial: think of Bergson on film, Freud on the phonograph, or Gunther Anders on television. Sonic or visual characteristics now affectionately described as warm and organic were described as cold and mechanical (Pinch and Trocco 2002; Hilderbrand 2009a). And the senses themselves continue to have a history after the nineteenth century, where they are understood as consisting of discrete operations as often as they are understood of consisting of continuous (Mills 2011; Moore 2003). In other words, the harmony and universality of *the analog* is itself imaginable only under certain historical conditions: the media era we now call "the analog era" and the coterminous moment in the history of science when the senses—and reality itself—were imagined through wave metaphors. By the mid-twentieth century, both conditions were on the decline.

The idea that analog media are more like the senses or more accurately limn the world's workings requires some retrospective imagination. For instance, waves are a particularly loaded figure of speech for describing the world. As Tara Rodgers (2011) argues, the wave metaphor for sound is ancient in origin. Talk of sound as waves was not properly the domain of Newtonian physics; rather, Newtonian physics is one application of a set of metaphors with considerably richer cultural history, one in which acousticians would have been bound up. Rodgers shows that classic acoustics texts like those of John William Strutt (Baron Rayleigh) and John Tyndall made use of common maritime themes of exploration, discovery, and control, as well as classic modernist tropes of masculine mastery over feminine nature. In the twentieth century, maritime figures shaped the description of sound synthesis and signal processing technologies, in press releases, technical diagrams, and sometimes on the instruments themselves. Rodgers ends her chapter on the wave metaphor with an argument that there are more feminist ways to conceive of and represent waves. But no matter which approach we take to the description of waves, we are operating within what Donna Haraway called situated knowledges (Rodgers 2011; Haraway 1991).

At the end of his discussion of the analog, Ted Friedman argues that the analog/digital binary is bivalent, and that scholars should instead think of reality in multivalent terms (2005, 45). That is sound advice, and we can now extend it. We should return some specificity to *the analog* as a particular technocultural sphere. That is to say that reality is just as analog as it is digital; and conversely, that it is just as not-digital as it is not-analog. Ultimately this goes back to an old argument, one made well by the last generation of technology scholars, ranging across methodological and political orientations, including Kittler and Massumi at other points in their writings: technology is part of the domain of human existence, not something outside it. The meanings we commonly attribute to the word *analog* did not even fully exist in the so-called analog era. Restoring some specificity to the term will help stimulate our technological imaginations (Balsamo 2011), and free us from the burden of a history that was only recently invented. But it will also introduce a new set of problems, for even a restricted conception of *the analog* is slippery and proliferates. Analogy is everywhere, central to a host of disciplines and countless aesthetic traditions. But instead of seeing the *analog* as life itself, if we understand it as a dimension of life, we can also restore its descriptive and analytical power, all the while also forcing ourselves to develop richer and more varied histories and theories of digital media.

What then for media theory? I would hope that this history makes it at least more difficult to map the analog/digital binary onto older binaries like present/absent, material/immaterial, and real/symbolic, to name three well-loved couplets. These are cherished fallbacks, but they actually push us away from some of the most important questions media theory can ask today: how meaning and collectivity work together; how symbols and technologies both define what it means to be human and how humans fit into the larger world, ethically, ecologically, politically, historically; and how we might live well in the large-scale societies we now inhabit. In the shadow of impending ecological catastrophe and ongoing violence and injustice, these are pressing questions. By shedding nostalgia for a past that was more inherently connected to nature, we free ourselves to imagine new ways—and revivify dormant and alternative traditions—of connecting nature, culture, history, and technology.

See in this volume: algorithm, archive, cloud, culture, digital, internet, prototype

See in Williams: capitalism, communication, culture, experience, man, media, nature, organic, romantic

Notes

1 Fully documenting the word's spread is beyond the scope of this entry. However, it is clearly present both in published books and in message boards, online forums, and journalism.

2 This argument expands on Wendy Chun's (2011, 59) critique of computer interfaces as "functional analogs to ideology *and* its critique" because they concretize imagined relations to invisible processes and structures.

3 There is also a connection between analog-to-digital conversion and the history of transduction, but that would be a keyword entry in itself (MacKenzie 2002; Helmreich 2007).
4 Here, I echo Lucas Hilderbrand's point that the term digital "extends to so many devices that claims to a singular aesthetic are difficult to justify" (2009b). I would add that it is also difficult to justify a singular ontology for all of them.
5 Thanks to Fred Turner for this background information.

References

Balsamo, Anne. 2011. *Designing Culture: The Technological Imagination at Work*. Durham, NC: Duke University Press.

Brand, Stewart. 1987. *Media Lab: Inventing the Future at MIT*. New York: Penguin Books.

Canales, Jimena. 2009. *A Tenth of a Second: A History*. Chicago: University of Chicago Press.

Chun, Wendy Hui Kyong. 2011. *Programmed Visions: Software and Memory*. Cambridge, MA: MIT Press.

Edwards, Paul. 1996. *The Closed World: Computers and the Politics of Discourse in Cold War America*. Cambridge, MA: MIT Press.

Friedman, Ted. 2005. *Electric Dreams: Computers in American Culture*. New York: New York University Press.

Goldstine, Herman H. 1993. *The Computer from Pascal to von Neumann*. Princeton, NJ: Princeton University Press.

Hankins, Thomas L., and Robert J. Silverman. 1995. *Instruments and the Imagination*. Princeton, NJ: Princeton University Press.

Haraway, Donna. 1991. *Simians, Cyborgs and Women*. New York: Routledge.

Helmreich, Stefan. 2007. "An Anthropologist Underwater: Immersive Soundscapes, Submarine Cyborgs and Transductive Ethnography." *American Ethnologist* 34(4): 621–41.

Hilderbrand, Lucas. 2009a. *Inherent Vice: Bootleg Histories of Videotape and Copyright*. Durham, NC: Duke University Press.

———. 2009b. "'Digital' Is Not a Noun." *FlowTV* 10:4 (July 23). http://flowtv.org/2009/07/"digital"-is-not-a-noun%C2%A0%C2%A0lucas-hilderbrand%C2%A0%C2%A0university-of-california-irvine%C2%A0/.

Jain, Sarah S. Lochlann. 2006. *Injury: The Politics of Product Design and Safety Law in the United States.* Princeton, NJ: Princeton University Press.

Kirschenbaum, Matthew. 2008. *Mechanisms: New Media and the Forensic Imagination*. Cambridge, MA: MIT Press.

Kittler, Friedrich. 1999. *Gramophone-Film-Typewriter*. Translated by Geoffrey Winthrop-Young and Michael Wutz. Stanford, CA: Stanford University Press.

Lacan, Jacques. 1998. *The Seminar of Jacques Lacan: The Four Fundamental Concepts of Psychoanalysis*. Translated by Jacques-Alain Miller and Alan Sheridan. New York: W. W. Norton.

Light, Jennifer. 1999. "When Computers Were Women." *Technology and Culture* 40(3): 455–83.

MacKenzie, Adrian. 2002. *Transductions: Bodies and Machines at Speed*. New York: Continuum.

Massumi, Brian. 2002. *Parables for the Virtual: Movement, Affect, Sensation*. Durham, NC: Duke University Press.

Mills, Mara. 2011. "Deafening: Noise and the Engineering of Communication in the Telephone System." *Grey Room* (43):118–43.

Moore, Brian C. J. 2003. *An Introduction to the Psychology of Hearing*. New York: Academic Press.

Pinch, Trevor, and Frank Trocco. 2002. *Analog Days: The Invention and Impact of the Moog Synthesizer*. Cambridge, MA: Harvard University Press.

Robinson, Derek. 2008. "Analog." In *Software Studies: A Lexicon*, edited by Matthew Fuller, 21–31. Cambridge, MA: MIT Press.

Rodgers, Tara. 2011. "Synthesizing Sound: Metaphor in Audio-Technical Discourse and Synthesis History." PhD diss., Art History and Communication Studies, McGill University, Montreal.

Sterne, Jonathan. 2003. *The Audible Past: Cultural Origins of Sound Reproduction*. Durham, NC: Duke University Press.

4

Archive

Katherine D. Harris

Digital, electronic, and hypertextual archives have come to represent online and virtual environments that transcend the traditional repository of material artifacts typically housed in a library or other institution (Price "Electronic" para. 3). Physical archives claim to amass anything that gives evidence of a time that has passed and "is essential for a civilized community" (*OED*). Traditional institutions define an archive as a rare book library, emphasizing the collecting of codex and manuscript representations of writing, authorship, and history. Most rare book and manuscript divisions also collect, preserve, and archive nontraditional forms of printed material and unconventional literary objects. This type of institutional archive is guided by principles of preserving history and an assumption that a *complete* collection will reveal not only that moment but also its beginning, ending, and connection to other moments. Voss and Werner articulate the duality of the archive as both physical space and, now, an imaginative site, both of which are governed by ideological imperatives (i).

Since approximately 1996, the *digital* archive has revised the traditional institutional archive to represent both a democratizing endeavor and a scholarly enterprise (Manoff 9). An archive, if truly liberal in its collecting, represents an attempt to preserve and record multiple metanarratives (Voss and Werner). Curators and special collection directors become editors and architects of digital archives to produce "an amassing of material and a shaping of it" (Price "Electronic" para. 9). However, the digital archive's instability threatens these metanarratives because of its potential for endless accumulation—a contamination.

In *Archive Fever*, Derrida suggests that the moments of archivization are infinite throughout the life of the artifact: "The archivization produces as much as it records the event" (17). Archiving occurs at the moment that the previous representation is overwritten by a new "saved" document. Traces of the old document exist but cannot be differentiated from the new. At the moment an archivist sits down to actively preserve and store and catalog the objects, the archive is once again contaminated with a process. This, according to Derrida, "produces more archive, and that is why the archive is never closed. It opens out of the future" (68). Literary works become archives not only in their bibliographic and linguistic codes,[1] but also in their social interactions yet to occur. It is the reengagement with the work that adds to an archive and that continues the archiving itself beyond the physical object.

My keyword essay follows a burning desire central to the archive—to return to the origins intermixed with the desire to hold everything at once in the mind's eye. In literature, this of course causes the protagonist to faint, go mad, isolate herself, create alternate realities—all in the name of either escaping or explaining what cannot be known. For example, my Gothic Novel students pointed out that the narrator in a short story, most specifically Lovecraft's, attempts to focus on a few actions in the busy-ness of the world, to focus the reader on what is calculable, knowable, but ultimately *unheimlich*, or the Freudian sense of uncanny.

In much the same way as a narrator, in the digital age we attempt to create archives of a particular moment or medium (The September 11 Digital Archive <http://911digitalarchive.org/>), the entirety of a medium (The Internet Archive <https://archive.org//>), the mutability of language (*The Oxford English Dictionary*), all knowledge (Wikipedia). More than others, the crowdsourced information of Wikipedia attempts to capture knowledge as well as the creation of that knowledge—the history or Talk of each Wikipedia entry unveils an evolving community of supposedly disinterested[2] users who argue, contribute, and create each entry. Wikipedia entries represent that digital version of an archive in the twenty-first century: the archive as a fractured, incalculable moment attempting to hold close all that happens at once in the world. But this concept has become incredibly problematic with the rush of information around us (Harris, "Archive," *The Johns Hopkins Guide to Digital Media*).

Kenneth Price begins my discussion about *archive* by offering a traditional definition of the term (see also **surrogate**):

> Traditionally, an archive has referred to a repository holding material artifacts rather than digital surrogates. An archive in this traditional sense may well be described in finding aids but its materials are rarely, if ever, meticulously edited and annotated as a whole. In an electronic environment, *archive* has gradually come to mean a purposeful collection of digital surrogates. ("Electronic" para. 3)

Later in this article, Price veers into discussing the role of archivist in shaping the archive; his description, though less dramatic, resembles Derrida's. Price's article is in response to the authority of a digital scholarly edition and its editors in the face of traditional print editions. Always, for Price, there is an organizing principle to archiving and, subsequently, editing. However, what we're concerned with in this keyword is inherently the messiness of the archive as it pertains to cultural records, both physical and digital. What gets placed into the archive and by whom becomes part of that record. What's missing, then, becomes equally important. Martha Nell Smith proposes that digital archives are free from the constraints of a traditional print critical edition; more importantly, the contents and architecture of a digital archive can be developed in full view of the public with the intention of incorporating the messiness of humanity.

In "Googling the Victorians," Patrick Leary describes all sorts of digital archives about Victorian literature that were springing up—archives that are not peer-reviewed per se but offer an intriguing and sanguine view of the wealth of nineteenth-century materials. Leary concludes his essay by asserting that whatever does not end up in a digital archive, represented as cyber- and hypertext, will not, in the future, be studied, remembered, valorized, and canonized. Though this statement reflects some hysteria about the loss of the print book, it is also revealing in its recognition that digital representations have become common and widespread, regardless of professional standards. Whatever is not on the web will not be remembered, says Leary (see also **memory**). Does this mean that the literary canon will shift to accommodate all of those wild archives and editions? Or does it mean that those mega projects of canonical authors will survive while the disenfranchised and noncanonical literary materials will fall further into obscurity?

Raymond Williams posits that "vulgar misuse" allows for entry into the cultural record (Williams *Keywords* 21), though those in library science object to a normalization of "archive" that moves away from their professional standards for a vault for the records of humanity. But the construction of a digital archive in literary studies conflates literature, digital humanities, history, computer programming, social sciences, and a host of other cross-pollinated disciplines. The archive, more than anything right now in literary studies, demonstrates what Williams calls "networks of usage" (23) with "an emphasis on historical origins [as well as] on the present—present meanings, implications, relationships—as history" (23). Community, radical change, discontinuity, and conflict are all part of the continuum in the creation of meaning according to Williams, seemingly similar to Borges's "Library of Babel" and Derrida's "archive fever." While archivists insist on a conscious choice in the use of *archive* (noun or verb), perhaps as part of a professional tradition, I seek to look at the messiness of the word as a representation of the messiness of our means for porting records across the past, present, and future.

The issue with formal digital archives is where to stop collecting to account for scope, duration, and shelf space. In digital archives, sustainability is key; but the digital archive is vastly more capable of accumulating everything and then allowing its holdings to be subjected to liberal and even promiscuous remixing by its users based on the tools available. The primary argument seems to concern who is controlling the inventorying, organization, tagging, coding of the data in service of an archive (user, curator, editor, architect?). And what digital tools are best employed in sorting the information? Even a tool offers a preliminary critical perspective.

For instance, in "The Master's Tools Will Never Dismantle the Master's House" (1984), Audre Lorde identified a schism in feminism that highlights missing voices, those voices that did not align themselves with patriarchal control, voices that refused to work within the system to gain power. In digital humanities' interactions with literary, library, and media studies, especially in the construction of databases, digital archives, and repositories, those marginalized voices exist, but they exist outside the scope of the traditional literary canon even still. Amy Earhart and Jamie Skye Bianco both notice this lack in digital representations of historical and literary materials; while Earhart focuses on the lack of diversity and the replication of the standard literary canon in "Can Information Be Unfettered? Race and the New Digital Humanities Canon," Skye Bianco asserts something more provocative about the very infrastructure of digital humanities:

> Boiled down blithely, the theory is in the tool, and we code tools. Clearly this position never refers to Audre Lorde's famous essays on tools nor to "the uses of anger," but it does summon their politics. . . . Tools don't reflect upon their own making, use, or circulation or upon the constraints under which their constitution becomes legible, much less attractive to funding. They certainly cannot account for their circulations and relations, the discourses and epistemic constellations in which they resonate. They cannot take responsibility for the social relations they inflect or control. Nor do they explain why only 10 percent of today's computer science majors are women, a huge drop from 39 percent in 1984, and 87 percent of Wikipedia editors—that would be the first-tier online resource for information after a Google search—are men. Tools may track and compile data around these questions, visualize and configure it through interactive interfaces and porous databases, but what then? What do *we do* with the data? (99)

The tools, like markup, by their very nature enact a sort of politics that replicates these archival silences that Lauren Klein discusses in reference to American slavery. By offering a "stable publication environment" and peer review to small-scale digital scholarly editions, the 2012 inaugural issue of the revised *Scholarly Editing*, under the editors Amanda Gailey and Andrew Jewell, attempts to balance the digital offerings of cultural materials beyond canonical authors and figures. But, in all of these scenarios, the relationship with the user is also absent; how do users shape an archive and how do those tools implemented by users reshape and remix that very same archive?

Kathleen Burnett, borrowing from Deleuze and Guattari's concept of the rhizome, notes in "Toward a Theory of Hypertextual Design" that the archive is less about the artifact and more about the user:

> Each user's path of connection through a database is as valid as any other. New paths can be grafted onto the old, providing fresh alternatives. The map orients the user within the context of the database as a whole, but always from the perspective of the user. In hierarchical systems, the user map generally shows the user's progress, but it does so out of context. A typical search history displays only the user's queries and the system's responses. It does not show the system's path through the database. It does not display rejected terms, only matches. It does not record the user's psychological responses to what the system presents. . . . *The map does not reproduce an unconscious closed in upon itself; it constructs the unconscious.* (25; italics added)

The digital archive, some argue, is the culmination of Don McKenzie's "social text," and the database, and to some extent hyperlinks, allow users to chase down any reference. In essence, the users become ergodic and radial readers. McGann, in *The Textual Condition*, defines radial reading as the activity of reading that regularly transcends its own ocular physical bases, which means that readers leave the page in order to acquire more information about the book (i.e., look up a word in the dictionary or flip to an endnote). This allows the reader to interact with the book, text, story through this acquisition of knowledge. The reader makes and remakes the knowledge produced by the text through this continual knowledge acquisition, yet the reader never actually leaves the text. It stays with her even while she consults other knowledge. This creates a plasticity to the text that is unique according to each reader (119).

But the archive, a metaphor once again, is always and forever contaminated, according to David Greetham in *The Pleasures of Contamination*. An archive is less about the text of a printed word and can be about all facets of materiality, form, and its subsequent encoding—even the reader herself. Scott Rettberg notes that the act of reading prioritizes the experience over the object itself with this idea of ergodic reading:

> The process of reading any configurative or "ergodic" form of literature invites the reader to first explore the ludic challenges and pleasures of operating and traversing the text in a hyperattentive and experimental fashion before reading more deeply. The reader of Julio Cortazar's *Hopscotch* must decide which of the two recommended reading orders to pursue, and whether or not to consider the chapters which the author labels "expendable." The reader of Milorad Pavic's *Dictionary of the Khazars* must devise a strategy for moving through the cross-referenced web of encyclopedic fragments. The reader of David Markson's *Wittgenstein's Mistress* or *Reader's Block* must straddle between competing desires to attend to the nuggets of trivia of which those two books are largely composed or to concentrate on the leitmotifs which weave them into a tapestry of coherent psychological narrative. In each of these print novels, the reader must first puzzle over the rules of operation of the text itself, negotiate the formal "novelty" of the novel, play with the various pieces, and fiddle with the switches, before arriving at an impression of how the jigsaw puzzle might together [*sic*], how the text-machine may run. Only after this exploratory stage is the type of contemplative or interpretive reading we associate with deep attention possible. (para. 13—emphasis added)

As our understanding of digital interruptions in an otherwise humanistic world expands and becomes both resistant and welcoming, we find that in these and other textual encounters, the definition of *archive* expands as well.

See in this volume: cloud, event, memory, prototype, surrogate

See in Williams: alienation, art, consensus, creative, empirical, history, institution, interest, management, representative, standards, subjective, tradition

Notes

1 The *bibliographic code* is distinguished from the content or the semantic construction of language within a text (linguistic code) by the following elements, as George Bornstein describes: "features of a page layout, book design, ink and paper, and typeface . . . publisher, print run, price or audience. . . . [Bibliographic codes] might also include the other contents of the book or periodical in which the work appears, as well as prefaces, notes, or dedications that affect the reception and interpretation of the work" (30, 31). *Linguistic codes* are specifically the words. Also within the book are paratextual elements that do not necessarily fall under the bibliographic or linguistic codes.

2 See Matthew Arnold on disinterestedness in *Essays on Criticism* <http://www.bartleby.com/223/0407.html>.

References

"Archive." *OED* 1959 Chambers's Encycl. I. 570/1.

Bianco, Jamie Skye. "This Digital Humanities Which Is Not One." *Debates in the Digital Humanities*. Ed. Matthew Gold. Minneapolis: University of Minnesota Press, 2012.

Bornstein, George. "How to Read a Page: Modernism and Material Textuality." *Studies in the Literary Imagination* 32:1 (Spring 1999): 29–58.

Burnett, Kathleen. "Toward a Theory of Hypertextual Design." *Postmodern Culture* 3:2 (January 1993): 1–28.

Derrida, Jacques. *Archive Fever: A Freudian Impression*. Trans. Eric Prenowitz. Chicago: University of Chicago Press, 1996.

Earhart, Amy. "Can Information Be Unfettered? Race and the New Digital Humanities Canon." *Debates in the Digital Humanities*. Ed. Matthew Gold. Minneapolis: University of Minnesota Press, 2012. 309–18.

Gailey, Amanda, and Andrew Jewell. Introduction. *Scholarly Editing: The Annual of the Association for Documentary Editing* 33 (2012). <http://www.scholarlyediting.org/2012/essays/essay.v33intro.html>.

Greetham, David. *The Pleasures of Contamination: Evidence, Text, and Voice in Textual Studies*. Indiana University Press, 2010.

Harris, Katherine. "Archive." *The Johns Hopkins Guide to Digital Media*. Eds. Marie-Laure Ryan, Lori Emerson, and Benjamin J. Robertson. Baltimore: Johns Hopkins University Press, 2014.

Klein, Lauren F. "The Image of Absence: Archival Silence, Data Visualization, and James Hemings." *American Literature* 85:4 (2013): 661–88.

Leary, Patrick. "Googling the Victorians." *Journal of Victorian Culture* 10:1 (Spring 2005): 72–86.

Lorde, Audre. "The Master's Tools Will Never Dismantle the Master's House." 1984. <http://collectiveliberation.org/wp-content/uploads/2013/01/Lorde_The_Masters_Tools.pdf>.

Manoff, Marlene. "Theories of the Archive from across the Disciplines." *portal: Libraries and the Academy* 4:1 (2004): 9–5.

McGann, Jerome. "How to Read a Book." *The Textual Condition*. Princeton, NJ: Princeton University Press, 1991. 119.

Price, Kenneth. "Edition, Project, Database, Archive, Thematic Research Collection: What's in a Name?" *Digital Humanities Quarterly* 3:3 (Summer 2009) <http://digitalhumanities.org/dhq/vol/3/3/000053/000053.html>.

———. "Electronic Scholarly Editions." *A Companion to Digital Literary Studies.* Eds. Susan Schreibman and Ray Siemens. Oxford: Blackwell, 2008. <http://www.digitalhumanities.org/companionDLS/>.

Rettberg, Scott. "Communitizing Electronic Literature." *Digital Humanities Quarterly* 3:2 (Spring 2009). <http://www.digitalhumanities.org/dhq/vol/3/2/000046.html>.

Smith, Martha Nell. "The Human Touch: Software of the Highest Order." *Textual Cultures* 2:1 (Spring 2007): 1–15.

Voss, Paul, and Marta Werner. Introduction. "Towards a Poetics of the Archive." *Studies in the Literary Imagination* 32:1 (1999): i–vii.

Williams, Raymond. *Keywords: A Vocabulary of Culture and Society*. New York: Oxford University Press, 1983.

5

Cloud

John Durham Peters

Nothing is quite so packed with meaning as clouds. Despite their reputation as flighty and insubstantial, clouds have carried a wide range of discourses, practices, and arts for a very long time. Perhaps it is precisely their apparent blankness, mutability, and vanishing mode of being that makes them such a ripe canvas for human creativity and criticism. The term *cloud* has earned its place as a digital keyword because of its widespread recent use for server-based online data storage, but the nearly instant and universal acceptance of this term would be impossible without the much longer legacy that I trace here.

Cloud is etymologically related to *clod*, and the *Oxford English Dictionary* reports that the first but now obsolete meaning of *cloud* was rock or hill. Perhaps the atmospheric rather than geological sense of the term, emerging in thirteenth-century English, was originally a metaphorical projection of terrestrial to celestial cumulus quite like the more recent usage. Tracing the comparative pathways of cognates shows how closely sky and clouds have been associated. *Sky* in Norwegian, a cognate to the identically spelled English term, actually means cloud, and *welkin* in English, an archaic term favored by Shakespeare and other poets that means sky or celestial vault, is cognate with the Dutch *wolk* and the German *Wolke*, both of which actually mean cloud. The deep association of clouds with the upper sphere or celestial realm has conditioned much of the history of their meaning. Clouds can also be found in deep space—for example, the Magellanic Clouds (dwarf galaxies) visible in the Southern Hemisphere, or interstellar clouds of dust and gas known as *nebulae*, from the Latin word meaning clouds. To be "in the clouds" has long meant to be in the sky, and by implication, to be in a fanciful, mystical, or "ungrounded" state. In the digital "cloud," the sense of whim, instability, or risk seems remarkably absent, perhaps enforced by the consistent use of the singular "in the cloud" to contrast with "in the clouds."

Throughout its varying history *cloud* has always meant an agglomeration or amassing of materials, whether of stone, water vapor, or data. Thus the koinē Greek νέφος μαρτύρων "cloud of witnesses" (Heb. 12:1) was solidified in English via the King James Bible (1611). In *Paradise Lost* (1667) John Milton mentions a cloud of locusts, and, most intriguingly, the *OED* supplies the 1705 phrase "a cloud of informations." Clouds could be crowds or swarms of arrows, flies, or birds, anything bunched that can cast a shadow. This sense also extends to use of *cloud* as a verb in English since the sixteenth century. *To cloud* means to cover with darkness, obscure, or dim, and can have the related sense of ill humor or gloom, as in "cloud of suspicion" or "under a cloud." Such negative meanings, remarkably, have little currency in the digital *cloud*.

Perhaps the oldest discourse around clouds is a theological one. In Homer νεφεληγερέτα is an epithet for Zeus ("he who collects the clouds"); clouds can be just as important attributes of divinity as thunder and lightning. In the Hebrew Bible, YHWH has his habitation in "the cloud" (not clouds) and guides the people of Israel on their desert sojourn with a pillar of cloud, a sign that both obscures his presence and also thereby points to it. (This usage foreshadows the recent notion that the cloud is secure and infallible.) In the New Testament, Christ is said to return to the earth "in the clouds," a theme beloved of baroque painters; in popular culture, there are innumerable images of a cloud heaven inhabited by God, saints, and angels. Owing to their association with the celestial realms, clouds are ready metonyms for deities of all kinds. Cross-culturally, deities associated with clouds are not always male: in Norse mythology, Frigg is a goddess who spins the clouds and tells the future, and in China, "the play of clouds and rain" can be a metaphor for sexual union.[1] I can hardly begin to catalog all the things that clouds have meant.

A counterdiscourse around clouds is meteorological, that is, reading clouds for physical rather than metaphysical signs. Farmers and sailors have known that clouds are harbingers of weather, and Aristotle uses the sky to make an apparently banal point about interpretation in the *Rhetoric*: "if it is cloudy, it will probably rain."[2] Though it is difficult, as Friedrich Kittler points out, to separate weather and the gods, there have been efforts since antiquity to read the sky secularly and scientifically.[3] In *The Nature of Things*, Lucretius reproved people who found faces and animals in the clouds, saying that we should see them as the fortuitous motions of atoms.[4] Other ancient thinkers emphasized the random quality of cloud shapes and taught us instead to understand images in the clouds as figments of imagination and thus as resting entirely in the viewing subject and not in the nebular object. (In *Calvin and Hobbes*, Calvin spots a cloud and says, "It must be a sign!" "Of what?" asks Hobbes. "Of very peculiar high altitude winds, I guess.") This debunking reading of clouds is antitheological in the case of Lucretius and antiphilosophical in the case of the comic playwright Aristophanes, whose play *The Clouds* mocks Socrates, his head-in-the-clouds thoughts, and "Cloud Cuckoo Land." Both authors associate clouds with airy, theoretical, insubstantial things, whether gods or "ideas."

There are elements in Jewish and Christian religion that are just as critical of reading God or anything else directly in the clouds. The Hebrew prophets were fiercely iconoclastic and denounced the reading of clouds for portents or omens, and God stumps Job by asking him to explain the origin of clouds, as if this were something that would forever elude his understanding. Jesus, squarely in this tradition, when asked to show some curious spectators a sign in the heavens, gave an answer as sarcastic as anything in Lucretius. If you want a sign in the sky, he said, here is one: red sky at night means good weather tomorrow, and red sky at morning means bad weather. His "sign" was a lesson in everyday forecasting.[5] Those who say there are no gods in the sky or say there is only one beyond it can agree that it is foolish to look for divine signs in the clouds; the antidivination discourse about clouds has both atheistic and (mono)theistic sources. Yet in popular religiosity the clouds remain something to conjure with, habitations for all manner of heavenly entities.

The debunking discourse around clouds comes down especially hard on the practice of finding animal and human shapes in them. Hamlet's toying with Polonius must be the most famous example:

> **HAMLET:** Do you see yonder cloud that's almost in the shape of a camel?
> **POLONIUS:** By th' mass and 'tis, like a camel indeed.
> **HAMLET:** Methinks it is like a weasel.
> **POLONIUS:** It is backed like a weasel.
> **HAMLET:** Or like a whale.
> **POLONIUS:** Very like a whale. (*Hamlet*, 3.2.361–67)

Polonius's foolish suggestibility, as mutable as the clouds themselves, enforces the idea that clouds completely lack objectivity, a notion found in the use of the cloudlike "thought bubble" in cartoons, a convention that links the mental privacy of subjective states with clouds. Here clouds are seen as the embodiment of everything flighty and thus inspire comic commentary on how people manage to create meaning out of the blue. Since anybody can find anything in them, clouds stand for the unreliable treachery of perception and for the whimsical unreliability of subjectivity itself. Thus some scientists call humanistic research "cloudy" or "sky-writing." Nonetheless, the British Cloud Appreciation Society has published a coffee-table book and maintains a website dedicated to charming and droll images of "clouds that look like things"; humorous routines about what clouds look like remain a staple in popular culture.[6]

One of the standard jokes in the that-cloud-looks-like-a-? repertoire is "that cloud looks like a cloud." In fact, there is a strong tradition in the past five centuries of looking intently at and reading clouds for their own sake in art, media, and science. Take art first. Clouds have been a repeated and prominent subject in European painting since at least the Italian Renaissance. The puzzle is why clouds would proliferate when they seem to defy the revolutionary technique of linear perspective, which so changed Western painting, drawing, and architecture. Clouds lack clear edges, morph rapidly, exist as much in color as in shape, and are not exactly objects like other objects. This is the question art historian Hubert Damisch explores in his great book on cloud painting.[7] Damisch sees in their difficulty a chance for painters to both defy the geometric strictures of perspective and show off their skills. Clouds—abstract, aniconic, sheer image without likeness—are the "other" to linear perspective. A cloud painter such as the seventeenth-century Dutch master Jacob van Ruisdael does not see camels, weasels, and whales in the sky: rather he rigorously documents the clouds in all their visual glory. Meteorologists even claim to be able to find in his paintings reliable historical testimony of weather patterns.[8] In cloud painting artists depicted a curious kind of image that was not exactly symbol, icon, or index, but rather atmosphere and process, like the act of painting itself. (There are as many clouds as there are painters: clouds can offer hospitality to a wide range of styles without ceasing to look like clouds.)

In painting clouds were harbingers of a new kind of image, an abstract one of flow and turbulence rather than symbolic representation. They were among the first abstract objects to be depicted, and in this they are a critical early step in the history of recording media. Friedrich Kittler has famously argued that the acoustic and optical analog media of the nineteenth century caused a critical historical rupture: with photography, phonography, and cinema, the realm of the recordable expanded drastically to include nonintelligible and time-based objects, breaking writing's historical hold as the only medium of storage for any form of art or intelligence. "White noise" (*Rauschen*) was Kittler's preferred term for this new class of recordables, though this is also perhaps what cloud painters had been depicting for much longer. Writing could record the words "he sneezed" but never the complex motion or sound of sneezing, something that became routine in the later nineteenth century. Sound, motion, flow, process all became recordable and thus subject to time-axis manipulation and to analysis.[9]

The scientific standing of clouds benefited from these transformations. As historians of science have noted, sound recording and motion pictures should be contextualized within a wider range of nineteenth-century scientific instruments with names ending in *-scope* and *-graph* that allowed observation and inscription of temporal processes of all kinds, ranging from blood pressure and weather to noise and heat. The "graphic method," as French physiologist and proto-cinematographer Étienne-Jules Marey called it, followed innovations in mathematics and modeling—for example, Fourier equations—that could represent fluid dynamics such as sound, heat, and atmospheric aerosols (i.e., clouds). New methods of recording brought new objects onto the scientific agenda: heat, noise, smoke, glaciers, clouds. Indeed, the nineteenth century opened with the 1802 announcement by the British gentleman scientist Luke Howard of the standard nomenclature for cloud types. Though rivals proposed alternatives, his Latin-based classification of stratus, cumulus, cirrus, and their varieties has held steady.[10] Less successful were international scientific efforts starting in the late nineteenth century to assemble a "cloud atlas" made up of photographic images of the various types. The particularities of clouds defied the demand for standardization that an atlas requires.[11] Clouds are never quite capturable by language or thought; John Ruskin, the patron saint of the Cloud Appreciation Society, who wrote eloquently of cloud beauty in modern painting, said that a cloud is a mix of something and nothing.[12] This mixed ontological status was one of the things that so fascinated nineteenth-century writers and artists such as Goethe and Emerson, Shelley and Baudelaire, Constable, Turner, and Monet. The sea of faith might have been draining in the nineteenth century, as Matthew Arnold said, but many still took comfort in the clouds as heavenly objects.[13]

A history of cloud science in the last two centuries would also be a history of our media technologies. Around the turn of the twentieth century, the Scottish physicist C.T.R. Wilson invented the cloud chamber. Seeking at first to model cloud and water vapor formation, he hit upon an instrument that, thanks to its ability to detect and trace subatomic particles, played an essential role in particle physics in the first half of the twentieth century. If clouds were once seen as cloaks of reality, in the cloud chamber they had become its revelators.

A similar path between obscuring and revealing was traveled with satellites and computers. During the Cold War, high-altitude spy planes could take pictures of nuclear facilities and other targets on the ground only if there was no cloud cover. One impetus for weather forecasting was espionage: how to predict when clouds would part enough to justify a dangerous surveillance mission. Clouds may have blocked the intelligence cameras, but they revealed a great deal about weather to those who learned to read their patterns. Though some had looked down on clouds from mountaintops before, one revelation from twentieth-century high-altitude aerial photography is that our planet is covered with clouds.[14] Tracking and modeling clouds requires enormous amounts of data. Though the standard story of modern computer science emphasizes the desire to model nuclear explosions and their aftermath, the demand for weather data has been just as formative in advancing digital technology. (John von Neumann, for instance, was just as passionate about computer applications for weather as about those for nuclear explosions.) The first world wide web was arguably formed for watching weather, well before the web as we know it. Clouds sit at the heart of crucial innovations.[15]

Fractal geometry was another spin-off of cloud study, which encouraged both a logic of vagueness and the analysis of the indefinite. Philosophers say that clouds illustrate the "sorites paradox" or heap problem: two grains of rice do not make a heap, and you can remove two grains from a heap without its ceasing to be a heap, but through the adding or subtracting of grains, a heap will either come or cease to be. There is a vague boundary between heap and nonheap that can never be numerically specified. In the same way, there are many possible surfaces that can plausibly claim to be the edge of a cloud.[16] Ontological indefiniteness is part of the cloud's great intellectual fascination.

It is easy to say that clouds do not mean anything, but the deeper fact is that clouds mean a great deal and that the collective future of the human species may depend on reading them well, at least if we think about the rising anthropogenic concentration of atmospheric carbon and the radical changes to climate it implies. Now we face clouds that are no longer undisturbed natural artifacts. Smokestacks, nuclear bombs, cloud seeding, and geoengineering schemes show that many crucial clouds are artificial. The artifactual character of clouds is emphasized by recent artists such as Fujiko Nakaya, who builds cloud and mist installations, Berndnaut Smilde, who creates and photographs surrealist clouds *inside* of buildings, or "Monsieur Moo," who, in a performance about the legal ownership of clouds, transported rain-cloud balloons across the US-Canada border in civil disobedience of international law. (There are treaties against cloud-manipulation, e.g., causing flood or drought for military purposes.) Clouds are the exact sort of things that Bruno Latour likes to call "hybrids" or "imbroglios."

The first result on a Google search for *cloud* I get is an ad for "Microsoft Cloud," with an image of a data center topped by a puffy cumulus, as if Microsoft benefited from a celestial benediction. This use of the term *cloud* may have started in engineering diagrams of networks, but it almost instantly took to the sky, taking selective advantage of the surplus and residue of the term. The web is full of beatific images of laptops sitting on heavenly clouds. The rhetoric of the data cloud likes to exploit the peaceful, inconsequential parts of the tradition of cloud meanings while suppressing the rest. "The cloud" is a huge PR achievement for the IT industry, but it is profoundly deceptive.[17] For one thing, "the cloud" of online storage is neither natural nor environmentally friendly: it consists of a gigantic infrastructure of data centers, and the worldwide IT electricity use is estimated to equal that of Japan and Germany combined.[18] For another, the notion of "the cloud" downplays the risk of giving up control over our data. We might think about the term *cloud-attack* from World War I (a barrage of poison gas) or *cloudburst* from meteorology to counter the blithe IT ideology of the cloud. To entrust our data to "the cloud" may invoke old ideas of the benevolent gods above, but the more interesting part of the history of clouds is how much human-built meaning is there to be exploited if you know how to do so. The IT industry would like us to recall nothing but cloud illusions, as Joni Mitchell sang, but in this case it is better to try to know clouds from both sides now. In all moments of history, this would be the worst to think of clouds as purely immaterial, natural, and meaningless things.

See in this volume: archive, culture, flow, internet, mirror, sharing

See in Williams: aesthetic, civilization, determine, ecology, experience, image, industry, mediation, myth, nature, organic, radical, romantic, science, status, taste, unconscious

Notes

1 Thanks to Rasmus Kleis Nielsen and to Guobin Yang for cross-cultural information.

2 Aristotle, *Rhetoric*, 1393a.

3 Friedrich Kittler, *Musik und Mathematik* 1:1 (Munich: Fink, 2006), 79.
4 *De rerum natura*, bk. 4, lines 166ff.
5 Matt. 16:2–3.
6 For a large and diverse list of examples of cloud gazing in popular culture, see http://tvtropes.org/pmwiki/pmwiki.php/Main/ThatCloudLooksLike.
7 Hubert Damisch, *Théorie du nuage. Pour une histoire de la peinture* (Paris: Seuil, 1972). Translated by Janet Lloyd as *A Theory of /Cloud/* (Stanford, CA: Stanford University Press, 2002).
8 Franz Ossing, "Haarlem's Crown of Clouds: Meteorology in the Paintings of Jacob van Ruisdael," trans. Kari Odermann, http://bib.gfz-potsdam.de/pub/wegezurkunst/haarlem_ruisdael_en.pdf. See also Werner Busch, "Wolken zwischen Kunst und Wissenschaft," in *Wolken: Welt des Flüchtigen*, ed. Tobias G. Natter and Franz Smola (Ostfildern: Hatje Cantz Verlag, 2013), 16–26.
9 *Gramophone, Film, Typewriter*, trans. Geoffrey Winthrop-Young and Michael Wutz (1986; Stanford, CA: Stanford University Press, 1999).
10 See Richard Hamblyn's highly readable but slightly hagiographic *The Invention of Clouds: How an Amateur Meteorologist Forged the Language of the Skies* (London: Picador, 2001).
11 Lorraine Daston, "The Science of Clouds, or: The Limits of Representation," University of Oslo, September 13, 2012.
12 John Ruskin, *Modern Painters*, vol. 5 (Sunnyside, UK: George Allen, 1888), 108.
13 On clouds in nineteenth-century thought and art, see Kurt Badt, *Wolkenbilder und Wolkengedichte der Romantik* (Berlin: de Gruyter, 1960); André Weber, *Wolkenkodierungen bei Hugo, Baudelaire, und Maupassant im Spiegel des sich wandelnden Wissenshorizontes von der Aufklärung bis zur Chaostheorie* (Berlin: Frank und Timme, 2012), and *Wolken: Welt des Flüchtigen* (see n. 8 above).
14 Robin Kelsey, "Reverse Shot: Earthrise and Blue Marble in the American Imagination," *New Geographies 4: Scales of the Earth*, ed. El Hadi Jazairy (Cambridge, MA: Harvard University Press, 2011), 10–16.
15 The whole paragraph is based on Paul N. Edwards's excellent *A Vast Machine: Computer Models, Climate Data, and the Politics of Global Warming* (Cambridge, MA: MIT Press, 2010).
16 http://plato.stanford.edu/entries/problem-of-many/.
17 See Vincent Mosco, *To the Cloud: Big Data in a Turbulent World* (Boulder, CO: Paradigm, 2014).
18 Mark P. Mills, "The Cloud Begins with Coal," http://www.tech-pundit.com/wp-content/uploads/2013/07/Cloud_Begins_With_Coal.pdf?c761ac.

6

Community

Rosemary Avance

The digital era poses new possibilities and challenges to our understanding of the nature and constitution of community, long a highly moralized and politicized keyword in the field of communication and elsewhere. Its historic uses range from a general denotation of social organization, district, or state; to the holding of important things in common; to the existential togetherness and unity found in moments of *communitas*. Our English-language *community* originates from the Latin root *communis*, "common, public, general, shared by all or many," which evolved into the fourteenth-century Old French *comunité* meaning "commonness, everybody." Originally the noun was affective, referencing a quality of fellowship, before it ever referred to an aggregation of bodies or souls. Traditionally the term has encompassed our neighborhoods, religious centers, and nation-states—historically, geographic and temporal birthrights, subjectivities unchosen by the individual. Today, we speak of a global community, made possible by communication technologies, and our geographically specific notions of community are disrupted by the possibilities of the digital, where physically distant, disembodied beings create what they also call community. But are the features and affordances of digital community distinct from those we associate with embodied clanship and kinship?

Philosophers and social scientists have long placed community at the apex of human association as a utopian model of connection and cohesion, a place where human wills unite for the good of the group. In Western thought, premodern notions of community are linked with singleness of reason (as in Socrates's philosophical tradition) and with singleness of purpose (as in Christian theological traditions). Early sociologists continued the tradition; in

1887, Ferdinand Tönnies[1] identified *Gemeinschaft* and *Gesellschaft*, roughly "community" and "society," as two ways of organizing social ties. Tönnies conceived of the community as a traditional structure wherein skills and trades, agrarian lifestyles, and ethnic and religious ties united individuals with what Durkheim called a collective consciousness, where the contents of an individual's consciousness were largely held in common with the group.[2] For Durkheim, modernity's best hope rests in a reclaiming of the collective consciousness, which is threatened by the division of labor and the rise of the individual.

In postmodernity, the very concept of community is rife with sacred implications, redeeming the isolated individual from a society in which "forms of life are dislocated, roots unsettled, traditions undone . . . at sea in a world where common meanings have lost their force."[3] Victor Turner[4] adopts the source of the word in his theories of liminality and *communitas*, arguing after van Gennep that ritual rites of passage move an individual from a state of social indeterminacy to a state of communal oneness and homogeneity. The outcome of a liminal individual's reincorporation into a group is a burdening, as the individual takes on obligation and responsibility toward defined others. This is the formation, the very root, of community—an ethical orientation outside oneself and toward others. The implication of community, then, is citizenship-belonging. Community results when individuals accept and serve their obligations and responsibilities vis-à-vis the collective—a normative model exported to the West and yet built on fieldwork done in a traditional Ndembu village.[5]

Thus privileging a traditional community model formed around family, clan, or tribal affiliations, this Western obsession springs from a postmodern malaise around community's purported demise in an industrialized and urbanized landscape.[6] Paradoxically, modern mass communication technologies have long played a leading role in the struggle to regain a semblance of premodern, pretechnological community. John Dewey argued in 1927 that journalism could reproduce the effects of traditional communities, uniting a public into a community—a collective of individuals who emerge in consequence of their relation one to the other.[7] Importantly, Dewey's concern—like concerns about digital communication today—centered on a lack of face-to-face communication in industrialized life and proposed a mediated cure. He believed mass communication could reinstate consensus in social life. Likewise, Robert Park, working with the Chicago School of sociologists, argued that despite the dangers of consumerism, party politics, and petty gossip (prescient technological concerns, these), the role of the newspaper was to "reproduce, as far as possible, in the city the conditions of life in the village."[8]

Nearly a century later, we continue to place our hopes in mediation technologies to regain some imagined pure past, but with an ambivalence that permeates our understanding of what technology is and what it can do. As media theorist Felicia Wu Song has noted, "To study virtual communities is to delve into questions about our cultural beliefs about technology."[9] Despite our grand hopes, moral panics accompany all new media technologies, and the pronounced fear associated with global connectivity via the internet, with no little irony, reflects a long-standing fear of disconnection. Even before internet saturation, Robert Putnam notoriously gave voice to this fear in *Bowling Alone*,[10] suggesting that declines in community commitment manifest in low civic and political engagement, declining religious participation, increasing distrust, decreasing social ties, and waning altruism are at least in part attributable to technology and mass media, as entertainment and news are tailored to the individual and consumed alone. Putnam paints a bleak image of Americans in dark houses, television lights flickering and TV dinners mindlessly consumed. It is the dark side of liberal individualism, our modern anomie.

Digital community seems to offer a panacea to the fear of disconnection in a new media age. Framed as a solution to the problem of modernity, disembodied cyberspace somehow at once flattens and broadens our notions of self. Still, "digital community" is an elusive concept to pin down with precision. Some scholars differentiate between "virtual" and "digital" communities, the former denoting a quasi-geographical location (e.g., a virtual gaming community located at a particular URL), whereas digital communities are ephemeral, united around a shared interest or identity rather than a particular virtual location. But distinctions between "virtual" and "digital" communities remain fuzzy across disciplines, so that a reference to "online community" works as an imprecise but useful stand-in to refer to myriad forms of "relatively stable, long-term online group associations mediated by the internet or a similar network."[11] Various taxonomic models of online communities have been put forth, dividing them based on attributes, support software, relationship to the offline sphere, and boundlessness,[12] or based on community types such as interest, relationship, fantasy, and transaction.[13]

While past conceptions of community were generally outside one's agential selection—you are born and die in your town; your religion is the faith of your parents—today's diverse digital landscape means self-selection into communities of interest and affinity. But digital communities do not entirely escape the deterministic, as availability still marks a very real digital divide between those with access to the technology and those without. Not only that, but the affordances of various platforms, both in *intended* and *possible* uses, all inform what might be seen as a digital community's blueprint. Online community formation relies on this user-centered software architecture that predates the community itself, so that communities evolve and adapt not in spite of but because of the affordances of their technological platform. These include format, space constraints, visuals, fixity vs. mutability, privacy vs. surveillance, peer feedback, report features/terms of service, modality (cellular, tablet, desktop)—all features that inform what is possible in a given community. Digital communities can evade some but not all of the fixity of these structural constraints, reaching across a variety of platforms and forums on both the light and the dark web.

Many social networking sites like Facebook and Myspace are primarily "intentional communities" wherein self-selection into the platform and mutual "friending" secure one's place.[14] Other venues, like Twitter and blogs, involve asymmetric relationships, wherein one user may be "followed" without following back; these platforms may not in and of themselves be communities, but they can provide "the basis of interlinked personal communities" measured by membership, influence, integration/need fulfillment, and shared emotional connection.[15] Highly fragmented, niche communities redistribute power in both intangible and tangible ways—think only of the economic impact of peer-to-peer communities on the music industry, where file sharing challenges traditional conceptions of property rights and even our collective moral code.[16] Indeed, content sharing is the basis of online community—from photos, to text, to files and links—and users themselves decide their own level of engagement in these participatory cultures. Within the communities themselves, the flattening dynamic of internet culture, where everyone[17] can have a platform and a voice, obfuscates the very real social hierarchies that are supported by social processes and norms—all of which evolve from platform affordances.

Because the notion of community has been imbued with ethical and moral implications, claims that community can exist online are met with doubt, debate, and sometimes derision. Some scholars and observers express a reluctance to accept Facebook, Twitter, blogs, or forums as true examples of community, seeing these spaces as primarily narcissistic expressions of what Manuel Castells calls the "culture of individualism," emphasizing consumerism, networked individualism, and autonomy; rather than the "culture of communalism," rooted in history and geography.[18] Critical observers also rightly note that what is often called "community" online is in fact a consumer identity, created by marketers and sutured to users' understandings of their own behaviors.[19]

To be sure, not every site that calls itself a community is one—just as not every site that does not, is not. There are also functions and uses of cyberspace that do not involve or invoke community. Community, online as offline, is also a mixed bag: not all associations are good ones, and not all communities are effective for all users at all times. Yet some users, if we take them at their word, say that online community provides a space to be "real"—or somehow *more* authentic—in ways that offline, embodied community might sanction. An overabundance of narrative visibility and social support on the internet allows some users to foster difference in ways that limited offline social networks simply cannot sustain. That is to say, in today's world it is not uncommon for youth to self-identify as queer and first "come out" in digital spaces[20] or, to draw on my own ethnographic work, for Mormons to foster heterodox (e.g., liberal) identities in closed Facebook groups before what they too mark as a "coming out" to their conservative "real-world" family and friends.[21]

Lawrence Lessig has noted: "Cyberspace is a place. People live there. They experience all the sorts of things that they experience in real space, there. For some, they experience more. They experience this not as isolated individuals, playing some high tech computer game; they experience it in groups, in communities, among strangers, among people they come to know, and sometimes like."[22] When the placeness and spaceness of cyberspace become givens among scholars and observers, perhaps then the veracity of online community—as a possibility, and sometimes a reality—will be accepted by moralists and skeptics. We might do well to remember the origin of our term *community*, which referenced a quality of fellowship before it ever referred to an aggregation of souls. It seems our term has come full circle, as disembodied souls unite in fellowship mediated by the digital.

See in this volume: culture, event, forum, gaming, memory, sharing

See in Williams: city, class, collective, communication, communism, community, culture, ecology, ethnic, exploitation, individual, institution, mechanical, nationalist, native, organic, popular, society, technology, tradition

Notes

1 Ferdinand Tönnies, *Community and Association* [*Gemeinschaft und gesellschaft*], trans. C. P. Loomis (London: Routledge, 1955).
2 Emile Durkheim, *The Division of Labor in Society* (1893) (New York: Free Press, 2014).
3 Michael J. Sandel, "Morality and the Liberal Idea," *New Republic* 190(9) (May 7, 1984): 17.
4 Victor Turner, "Liminality and Communitas," in *The Ritual Process: Structure and Anti-Structure* (New York: Aldine, 1969), 94–130.
5 It should be noted that the valorization of the concept of community, while still normative, has not gone entirely unchecked. Structuralist theorists such as Michel Foucault and Benedict Anderson point out that the idealization of community as the ultimate social formation imbues it with a dangerous power. Foucault reminds us that pure community is at base an apparatus of control over social relations and interactions. Anderson reaffirms, too, the imaginary nature of community, which we conceive of as a "deep, horizontal comradeship" in such a way that power differentials are jointly pretended away.

6 Indeed, the discipline of sociology was established as a response to the perception of community's disintegration owing to industrialization and urbanization.
7 John Dewey, *The Public and Its Problems: An Essay in Political Inquiry* (University Park: Pennsylvania State University Press, 1927).
8 Robert E. Park, "The Natural History of the Newspaper," *American Journal of Sociology* 29(3) (1923): 277.
9 Felicia W. Song, *Virtual Communities: Bowling Alone, Together Online* (New York: Lang, 2009), 3.
10 Robert Putnam, *Bowling Alone: The Collapse and Revival of American Community* (New York: Simon & Schuster, 2000).
11 Andrew Feenberg and Darin David Barney, *Community in the Digital Age: Philosophy and Practice* (Lanham, MD: Rowman & Littlefield, 2004).
12 Jonathan Lazar and Jennifer Preece, "Classification Schema for Online Communities," *Proceedings of the 1998 Association for Information Systems Conference* (1998), 84–86.
13 John Hagel and Arthur G. Armstrong, *Net Gain: Expanding Markets through Virtual Communities* (Boston: Harvard Business School Press, 1997).
14 Cf. danah boyd, "Friends, Friendsters, and Myspace Top 8: Writing Community into Being on Social Network Sites," *First Monday* 11(12) (December 4, 2006). Available at http://firstmonday.org/article/view/1418/1336.
15 Gruzd Anatoliy, Barry Wellman, and Yuri Takhteyev, "Imagining Twitter as an Imagined Community," *American Behavioral Scientist* 55(10) (2011): 1294.
16 See Jerald Hughes and Karl Reiner Lang, "If I Had a Song: The Culture of Digital Community Networks and Its Impact on the Music Industry," *International Journal on Media Management* 5(3) (2003): 180–89.
17 "Everyone," that is, with access, equipment, technological savvy, and, presumably, an audience.
18 Manuel Castells, "Communication, Power, and Counter-power in the Network Society," *International Journal of Communication* 1 (2007): 238–66.
19 Cf. Robert V. Kozinets, "E-tribalized Marketing? The Strategic Implications of Virtual Communities of Consumption," *European Management Journal* 17(3) (1999): 252–64.
20 Mary L. Gray, "Negotiating Identities/Queering Desires: Coming Out Online and the Remediation of the Coming-Out Story," *Journal of Computer-Mediated Communication* 14(4) (July 2009): 1162–89.
21 Here I'm drawing on years of ethnographic work among Mormons on the internet, with details forthcoming in my dissertation, "Constructing Religion in the Digital Age: The Internet and Modern Mormon Identities"; for more on Mormon deconversion and online narratives, see Rosemary Avance, "Seeing the Light: Mormon Conversion and Deconversion Narratives in Off- and Online Worlds," *Journal of Media and Religion* 12(1) (2013): 16–24.
22 Lawrence Lessig, "The Zones of Cyberspace," *Stanford Law Review* 48(5) (May 1996): 1403.

7

Culture

Ted Striphas

Raymond Williams's *Keywords* isn't just a compendium of important terms in the English language. It's better imagined as a linguistic jigsaw puzzle, albeit one whose pieces are moving in relationship to one another. New terms get introduced, older ones drop out, and still others change shape as semantic edges grind together, altering the appearance of the whole. How else can one explain the unique structure of the work—each entry populated by a series of companion terms that, taken together, constitute a network of internal relations far more complex than anything suggested by the book's alphabetized table of contents? Or its iterative nature—beginning with the introduction to Williams's *Culture and Society* and its focus on the words *industry*, *democracy*, *class*, *art*, and *culture*, then mushrooming into the 110-entry first edition of *Keywords* published in 1976, and culminating in the revised edition of 1983, which added a further 21 terms (Williams 1958, xiii–xx; Williams 1976; Williams 1983a)? Little wonder Williams described *Keywords* as "necessarily unfinished and incomplete," having just performed a major overhaul (1983a, 27).

It is within this context that one ought to begin making sense of any keyword, including the one under consideration here, *culture*. First observation: in *Keywords* Williams says nothing explicit about *culture*'s relationship to digital technology. This isn't surprising given the historical ambit of the work, the endpoint for which is roughly the end of the third quarter of the twentieth century, when analog or at least predigital techniques still ruled the day (see **analog**). Second observation: the entry for *culture* is marked by its simultaneous distance from, and nagging referentiality to, the technological ethos of modern production. Williams states that *culture* "was used to attack what was seen as the 'MECHANICAL' character of the new civilization . . . emerging [in the nineteenth century]: both for its abstract rationalism and for the 'inhumanity' of current industrial development" (1983a, 89) Or, as Gilbert Simondon puts it: "while [culture] grants recognition to certain objects, for example to things aesthetic, and gives them their due place in the world of meanings, it banishes other objects, particularly things technical, into the unstructured world of things that have no meaning but do have a use,a utilitarian function" (1980, 2; cf. Horkheimer and Adorno 1997). Third observation: Williams saw fit to add *technology* to the 1983 edition of *Keywords*, an acknowledgment, perhaps, of the term's gathering import with respect to a vocabulary of culture and society. Yet, at three-quarters of a printed page, the write-up is barely a skeleton. (*Culture* gets six pages; *class*, the longest entry, nine.)

And so the trail linking *culture* to digital technology runs cold—unless one approaches the topic of keywords not in terms of a lone text but as an endeavor transecting multiple volumes in Williams's oeuvre. Moving outward from the ur-text, there emerges a Williams more attentive not only to technology (1974; 1983b), but also to the growing prevalence of the digital. To wit: in the penultimate section of his book *Culture*, published in 1981, he discusses the decline of industrial production in the West and, with it, the growth of "information processes."[1] He doesn't mention computational technologies by name, although he does refer to "data collection and processing" as activities integral to what Daniel Bell, one of Williams's interlocutors (1983a, 27), had termed "post-industrial society" (Bell 1973; see also Williams 1983b, 83–102). "Thus," writes Williams, "a major part of the whole modern labour process must be defined in terms which are not easily theoretically separable from the 'traditional' cultural activities" (1981, 231–32).

By 1981, then, Williams seems to have grasped how *culture* and *technology* were becoming less opposed than they once were, practically and theoretically. This was thanks in part to culture's budding relationship to digital information processing, a relationship mediated initially by large-scale institutional mainframes and, by the early- to mid-1980s, desktop personal computers. Fast-forward to the early 2000s, when these articulations have become so well established as to give rise to a host of lexical offshoots. The entry for *culture* appearing in the Williams-inspired reboot *New Keywords* mentions *cyberculture* and *technoculture* (Bennett 2005, 68); the volume also includes entries for *information*, *network*, and *technology* (Webster 2005a; Webster 2005b; Ross 2005). More recently Lev Manovich has employed the phrase *cultural software* "literally to refer to certain types of software that support actions we normally associate with 'culture'" (2013, 20).

It seems fair to say that a rapprochement between *culture* and *technology* has been achieved. But how has *culture*'s growing proximity to technology, particularly digital computational tools, affected the senses and meanings, the values and practices, with which the word is associated? "Culture," writes Simondon, "must come to terms with technical entities as part of its body of knowledge and values" (1980, 1). This is a matter of grammar, in the classical sense of apprehending words and then reconciling them both with and against the realities through which one moves.

Here it's useful to revisit the three understandings of *culture* Williams advances in *The Long Revolution*, published in 1961, which form a basis for the entry appearing in *Keywords*. Strictly speaking, these aren't so much definitions as "general categories" or rubrics under which Williams gathers a host of senses and meanings accessible at the time he was writing (1961, 57). They include the following:

1. the "ideal" definition, referring to the systems of valuation by means of which groups establish hierarchies, and subsequently judge the worth, of people, places, objects, institutions, and ideas;
2. the "documentary" definition, referring to the whole range of artifacts, both material and immaterial, produced by a group of people;
3. the "social" definition, referring to "a particular" or "whole way of life" (1961, 57; 1958, xviii), i.e., the patterns of thought, conduct, and expression, including the structures of signification, prevalent among members of a collective.

The entry for *culture* appearing in the *Oxford English Dictionary* traverses much the same ground as does *The Long Revolution*, suggesting that the array of senses and meanings Williams identified in his early work remain dominant reference points fifty years on ("Culture, n." 2014). But it's also worth bearing in mind the "sacral attitude" of dictionaries (Williams 1983a, 20), or their tendency to consecrate preferred usages at the expense of residual forms ("archaisms") and emergent ones ("vulgarizations").

Indeed, the story of *culture* post-1950 is centrally about archaism, or the reactivation of latent senses and meanings, and about vulgarization, or the appearance of novel understandings that seem to corrupt tried-and-true definitions of the word.

Culture doesn't exactly begin its career in the late eighteenth century, although it is around this time that the term leaves the semantic confines of husbandry and enters broader usage, gradually taking on the range of meanings encompassed by the three rubrics above (Williams 1983a, 87; see also Flusser 2013, 89–96). *Culture* then becomes a quintessentially modern term, carving out a conceptual space for human beings apart from nature on the one hand, and from technology on the other, subordinating both in the process (Latour 1993, 104).

The distinction from nature migrates into English primarily from German, establishing something like a mode of existence for human beings transcending the natural world. For example, in 1782 the influential German lexicographer Johann Christoph Adelung defined *culture* (*Kultur*) as "the transition from a more sensual and animal condition to the more closely knit interrelations of social life" (qtd. in Kroeber and Kluckhohn 1963, 37; see also McNeil 2005, 236; Marx 1964). His definition bespeaks the distance between *culture*'s modern form and its etymological taproot, the Latin *colere.* The latter denotes harvesting and cultivation, not in an instrumental sense but in a religious one, exemplified by the carryover into the English-language word *cult* (Flusser 2013, 90–91). *Colere* suggests the human species's dependency on and subordination to the natural world; *culture* loosens the tie and inverts the relationship (or at least gives the appearance of doing so).

The distinction from technology arises about a century later, mainly in England, fueled by the country's rapid industrialization and attendant concerns about the rising tide of proletarian democracy. The emergence of this sense is evident above all in Matthew Arnold's *Culture and Anarchy*, published in 1869. Arnold contended that *culture* was antithetical to industrialism, and more specifically to the machines out of which poured the run-of-the-mill, both literally and figuratively. "The idea of perfection"—his preferred view of *culture*—"is at variance with the mechanical and material civilization in esteem with us" (1993, 63; see also 78, 94). Thus Arnold championed the cause of public pedagogy, a pedagogy focusing on moral and spiritual development through exposure to impeccable art, literature, and other imaginative works whose instruments of production he refused to acknowledge. He had another aim too: to habilitate *culture*, which, owing to its connotation of pretentious learning, had hitherto played second fiddle to *civilization* in English-language usage.

So, in the nineteenth century, there emerges an overarching view of *culture* as "a court of human appeal" (Williams 1958, xviii), a view that aligns with the then-burgeoning phenomenological understanding of the lifeworld as an "autonomous realm" of human affairs (Kittler 2006, 42). This view is part and parcel of the birth of humanism, and of the humanities, the latter of which thematized *culture* and took it as its organizing motif (Williams 1983a, 150; Kittler 2006, 40–42). But *culture* doesn't shed its older, agricultural meanings completely. It retains a semblance of them in the Arnoldian belief, inherited from Johann Gottfried Herder, that *culture* consists of a long, deliberate process of nurturance and growth—although now selves are cultivated rather than soil and seeds.

These are decisive developments. In the near term they helped secure authority for the humanities, positioning both its practitioners and the disciplines to which they belonged as the leading arbiters of "cultural data" (Kittler 2006, 41). But in the long term they also helped precipitate a crisis, or rather a whole complex of crises that persist into the present day. Michel Foucault was among the most prescient observers of the coming troubles when, in 1966's *Les mots et les choses* (*The Order of Things*), he concluded:

> Man is not the oldest nor the most constant problem that has been posed for human knowledge. . . . It is not around him and his secrets that knowledge prowled for so long in the darkness. In fact, among all the mutations that have affected the knowledge of things and their order . . . only one, that

> which began a century and a half ago and is now perhaps drawing to a close, has made it possible for man to appear. . . . [M]an is an invention of recent date. And one perhaps nearing its end. (1970, 386–87)

Here, in his archaeology of the human sciences, Foucault glimpsed the beginning of the unraveling of modern humanism, a process that, by the closing decades of the twentieth century, would open *culture* to meanings, practices, and interpretive approaches that had largely been excluded for the better part of two centuries. Subsequent critics have suggested that "culture . . . has lost its purchase" as a result of these shifts (Readings 1996, 12). However, Lawrence Grossberg contends that *culture* remains a term of significance today, though "the ways in which it matters—and hence, its effects—have changed in ways that we have not yet begun to contextualize or theorize" (2006, 17).

Two puzzles, then: What has happened to humanism? And what is happening to *culture*, semantically, experientially, and theoretically? Donna J. Haraway and N. Katherine Hayles have gone further than most in addressing the former question. They identify the Second World War as a turning point when hermetic notions of "the human" began breaking down. For Haraway (1991) the shift is embodied in the figure of the cyborg and, for Hayles (1999), in that of the posthuman. While differing in important respects (Haraway 2006, 140), both figures trouble hard-and-fast distinctions between human beings, nature, and technology—the distinctions that helped secure the apparent autonomy of *culture* in the early nineteenth century. Moreover, Haraway and Hayles attribute the breakdown most immediately to the rise of cybernetics and information theory, many of whose key breakthroughs occurred within the context of the war (cf. Pickering 2010, 4). According to Haraway these fields provoked a "communications revolution," as well as a broader "re-theorizing of natural objects as technological devices properly understood in terms of mechanisms of production, transfer, and storage of information" (1991, 58).

The latter term—*information*—was the hinge on which this process swung. It functioned as a kind of counteranthropological leveler, an abstraction under which could be gathered a diverse array of expressive phenomena, both human and nonhuman (Schrödinger 1944, 70–71; Wiener 1954, 32; Bateson 2000, 272, 315–18; Peters 1988; Gleick 2011). The third quarter of the twentieth century saw a host of efforts to reconceptualize *culture* along these lines. Sociologist Talcott Parsons viewed it as an information-rich, cybernetic system (1970, 514–16), while his student, anthropologist Clifford Geertz, suggested that the operations of culture closely resembled those of computer software, given their shared concern for symbol processing (1973, 44). Parsons and Geertz were still operating at the level of analogy, however, viewing culture through a metaphorics of computation rather than positing an actual equivalence between them. Williams took it that next step in his claims about the entwining of cultural work and data processing. Tiziana Terranova has gone even further in suggesting that information now serves as a "milieu" or "environment within which contemporary culture unfolds" (2004, 8).

Here one might speak of the subsumption of *culture* under *information*, or rather its subsumption under the auspices of digital computational technologies. The term *subsumption* comes from Karl Marx, who uses it to identify two phases in the history of capitalist development. The first phase, or "formal" subsumption, refers to the capitalization of precapitalist relations, resulting in hybrid forms grafted on to the new mode of production. The second phase, or "real" subsumption, refers to the gradual emergence of properly capitalist productive relations, or relations that are capitalistic through and through (1976, 645–46, 1019–25). The history of *culture* in the second half of the twentieth century follows a similar trajectory, where formal analogies between culture and computation are now starting to realize themselves in a range of theories and practices that reconceptualize the former in terms of the latter. How else can one explain the emergence of mash-ups like "culturomics," the "digital humanities," and "humanities computing," or the reimagining of cultural artifacts as a corpus of "big data"?

One of the more intriguing outgrowths of all this has been a recognition, still dawning, of the ways in which *culture* exceeds human discourse, perception, and sense making. The work of Félix Guattari is exemplary in its insistence that human expression is but one element of an "assemblage of enunciation" whose ranks include "extra-linguistic, biological, technological," and other "a-signifying" modes of communicative practice (1995, 24). What one sees here is an awareness of how specific categories of signs, unintelligible to or unintended for humans, can nonetheless have a profound effect on the form, content, and delivery of culture. QR and other types of machine-readable product codes are a case in point (Striphas 2009, 81–110), as are techniques of search engine optimization, which "tune" websites for maximum discoverability by machines. What one also sees, then, is a stretching of the boundaries of culture beyond the "webs of significance" with which, in some formulations, it was once thought to be equivalent (Geertz 1973, 4). Given the degree to which machine-based systems now communicate about and process (sort, classify, prioritize) culture, it seems difficult to imagine it strictly as a "court of human appeal." One could reasonably see it as a court of machinic appeal as well (Hallinan and Striphas, in press).

All that is to say: sometime around 1950, the category *culture* starts to slide into the orbit of technology, having slipped, to a significant degree, the gravitational pull of modern humanism. With that an ostensibly antiquated sense of *culture*—the agrarian one referring to husbandry—is given a new lease on life. At first blush, the connections may not seem obvious. Computation seems to have little in common with "the tending of natural growth," *culture*'s original meaning in the English language (Williams 1983a, 87), also the sense both Herder and Arnold borrowed and twisted. Yet the semantic connections are there: in the notion of tending, indicating skill or technique, a derivative of the ancient Greek τέχνη (*technē*), from which the word *technology* derives (Stiegler 1998, 93); and also in *coulter*, a "subsidiary" form of the word *culture*, sometimes spelled as such, designating an instrument for tilling the soil or, as Nicholas John notes in this volume, for dividing and sharing (Williams 1983a: 87; see **sharing**). Once again, *culture* is becoming less distinct from its tools, and vice versa. Its story post-1950 thus exemplifies how "archaic" or residual forms press and persist, producing latencies of meaning that can reemerge under proper conditions.

To reiterate, this is not to suggest that the modern view of *culture*, exemplified by Williams's three definitional rubrics, is receding into the background. If anything the dominant view is compelled to cohabit with the emergent forms, producing what, in traditional lexicography, is apt to be understood as vulgarizations of meaning. Consider once again Matthew Arnold's *Culture and Anarchy*. The book proceeds from the assumption that "our social machine is a little out of order," and that *culture* is the "principle of authority" that will "counteract the tendency to anarchy" (1993, 88, 89). Despite Arnold's misgivings about modern technology, his view comports in an odd way with the position of applied information theory. Today, and to an unparalleled degree, Google and its kin adjudicate what Arnold once described as "the best which has been thought and said" (190). They do so by parsing signal and noise billions of times each day, in an effort to attenuate information overload. Though their means and ends differ, both Arnold and Google are invested in determining which aspects of human expression are most worthy of rising above the din. Both, therefore, are in the business of finding order amid the apparent chaos. Just as Arnold wrote *Culture and Anarchy*, so Google and company may well be writing the companion volume, *Culture and Entropy*.

Like any account of *culture*, this one, focusing on its relationship to digital technology, is partial—"necessarily unfinished and incomplete." This isn't a function of the focus, however, as much as it is a testament to the dynamism and adaptability of "one of the two or three most complicated words in the English language" (Williams 1983a, 87). Indeed, over the last fifty or sixty years *culture* has taken on new inflections—or rather reinflected older senses and meanings—many of which embody its current association with digital computational tools. The overview presented here thus is intended not as a narrow account of *culture*, circumscribed by a particular subject matter, but as one that significantly reflects the predicament of *culture* since the end of the Second World War.

See in this volume: algorithm, analog, cloud, democracy, digital, information, personalization, sharing

See in Williams: art, bureaucracy, civilization, common, communication, community, culture, generation, humanity, ideology, industry, machine, masses, media, nature, science, society, technology, western

Note

1 The US edition bears the title *The Sociology of Culture.*

References

Arnold, Matthew. 1993. *Culture and Anarchy and Other Writings.* Edited by Stefan Collini. Cambridge: Cambridge University Press.

Bateson, Gregory. 2000. *Steps to an Ecology of Mind: Collected Essays in Anthropology, Psychiatry, Evolution, and Epistemology.* Chicago: University of Chicago Press.

Bell, Daniel. 1973. *The Coming of Post-industrial Society: A Venture in Social Forecasting.* New York: Basic Books.

Bennett, Tony. 2005. "Culture." In *New Keywords: A Revised Vocabulary of Culture and Society,* edited by Tony Bennett, Lawrence Grossberg, and Meaghan Morris, 63–69. Malden, MA: Blackwell.

"Culture, n." 2014. *OED Online.* Oxford University Press.

Flusser, Vilém. 2013. *Natural:Mind.* Translated by Rodrigo Maltez Novaes. Minneapolis, MN: Univocal Publishing.

Foucault, Michel. 1970. *The Order of Things: An Archaeology of the Human Sciences.* New York: Vintage Books.

Geertz, Clifford. 1973. *The Interpretation of Cultures: Selected Essays.* New York: Basic Books.

Gleick, James. 2011. *The Information: A History, a Theory, a Flood.* New York: Pantheon Books.

Grossberg, Lawrence. 2006. "Does Cultural Studies Have Futures? Should it? (Or What's the Matter with New York?)." *Cultural Studies* 20(1): 1–32.

Guattari, Félix. 1995. *Chaosmosis: An Ethico-aesthetic Paradigm.* Translated by Paul Bains and Julian Pefanis. Bloomington: Indiana University Press.

Hallinan, Blake, and Ted Striphas. In press. "Recommended for You: The Netflix Prize and the Production of Algorithmic Culture." *New Media and Society.*

Haraway, Donna J. 1991. *Simians, Cyborgs, and Women: The Reinvention of Nature.* New York: Routledge.

———. 2006. "When We Have Never Been Human, What Is to Be Done?" *Theory, Culture & Society* 23(7–8): 7–8.

Hayles, Katherine. 1999. *How We Became Posthuman: Virtual Bodies in Cybernetics, Literature, and Informatics.* Chicago: University of Chicago Press.

Horkheimer, Max, and Theodor Adorno. 1997. "The Culture Industry: Enlightenment as Mass Deception." In *Dialectic of Enlightenment,* translated by John Cumming. New York: Continuum.

Kittler, Friedrich. 2006. "Thinking Colours and/or Machines." *Theory, Culture & Society* 23(7–8): 39–50.

Kroeber, A. L., and Clyde Kluckhohn. 1963. *Culture: A Critical Review of Concepts and Definitions.* New York: Vintage.

Latour, Bruno. 1993. *We Have Never Been Modern.* Cambridge, MA: Harvard University Press.

Manovich, Lev. 2013. *Software Takes Command.* New York: Bloomsbury.

Marx, Karl. 1976. *Capital: A Critique of Political Economy.* Translated by Ben Fowkes. 3 vols. New York: Penguin Press.

Marx, Leo. 1964. *The Machine in the Garden: Technology and the Pastoral Ideal in America*. New York: Oxford University Press.
McNeil, Maureen. 2005. "Nature." In *New Keywords: A Revised Vocabulary of Culture and Society*, edited by Tony Bennett, Lawrence Grossberg, and Meaghan Morris, 235–39. Malden, MA: Blackwell.
Parsons, Talcott. 1970. "Theory in the Humanities and Sociology." *Daedalus* 99(2): 495–523.
Peters, John Durham. 1988. "Information: Notes toward a Critical History." *Journal of Communication Inquiry* 12(2): 9–23.
Pickering, Andrew. 2010. *The Cybernetic Brain Sketches of Another Future*. Chicago: University of Chicago Press.
Readings, Bill. 1996. *The University in Ruins*. Cambridge, MA: Harvard University Press.
Ross, Andrew. 2005. "Technology." In *New Keywords: A Revised Vocabulary of Culture and Society*, edited by Tony Bennett, Lawrence Grossberg, and Meaghan Morris, 342–44. Malden, MA: Blackwell.
Schrödinger, Erwin. 1944. *What Is Life? The Physical Aspect of the Living Cell*. Cambridge: Cambridge University Press.
Simondon, Gilbert. 1980. *The Mode of Existence of Technical Objects*. Translated by Ninian Mellamphy. University of Western Ontario. https://english.duke.edu/uploads/assets/Simondon_MEOT_part_1.pdf.
Stiegler, Bernard. 1998. *Technics and Time: The Fault of Epimetheus*. Translated by Richard Beardsworth and George Collins. 3 vols. Stanford, CA: Stanford University Press.
Striphas, Ted. 2009. *The Late Age of Print: Everyday Book Culture from Consumerism to Control.* New York: Columbia University Press.
Terranova, Tiziana. 2004. *Network Culture Politics for the Information Age*. London: Pluto Press.
Webster, Frank. 2005a. "Information." In *New Keywords: A Revised Vocabulary of Culture and Society*, edited by Tony Bennett, Lawrence Grossberg, and Meaghan Morris, 186–89. Malden, MA: Blackwell.
———. 2005b. "Network." In *New Keywords: A Revised Vocabulary of Culture and Society*, edited by Tony Bennett, Lawrence Grossberg, and Meaghan Morris, 239–41. Malden, MA: Blackwell.
Wiener, Norbert. 1954. *The Human Use of Human Beings: Cybernetics and Society*. Rev. ed. Cambridge, MA: Da Capo Press.
Williams, Raymond. 1958. *Culture and Society, 1780–1950.* New York: Columbia University Press.
———. 1961. *The Long Revolution*. Orchard Park, NY: Broadview Press.
———. 1974. *Television: Technology and Cultural Form*. Hanover, NH: Wesleyan University Press.
———. 1976. *Keywords: A Vocabulary of Culture and Society*. 1st ed. New York: Oxford University Press.
———. 1981. *The Sociology of Culture*. Chicago: University of Chicago Press.
———. 1983a. *Keywords : A Vocabulary of Culture and Society*. Rev. ed. New York: Oxford University Press.
———. 1983b. *Towards 2000*. London: Chatto & Windus.

8

Democracy

Rasmus Kleis Nielsen

> Imagining the perfect democratic city does not exempt us from acting in the present scene of imperfection. . . . On the contrary, this imagining is what enables us to act, that is, to exist in freedom from a despair of democracy.
>
> —Stanley Cavell (1994)

If we want to understand the relationship between digital technology and democracy, we need to start with democracy. The word comes to English and many other languages via the Latin *democratia* from the ancient Greek δημοκρατία for popular government, composite of δῆμος (the commons, the people) and -κρατια (government, rule). It is often associated with classical Athens, though the Greeks had at best an ambivalent relationship with democracy, and the patriarchal, class-stratified, slave-owning city-state had no form of government we today would recognize as such. More broadly, historians have traced the roots of the idea of popular government back to "assembly democracies" first appearing in the region of what is today Syria, Iraq, and Iran, spreading to the Indian subcontinent, and then via the Phoenicians to the Greek world we today associate with the origins of democracy.[1]

Today, actually existing democracy is liberal and representative. It is liberal in being based on the rule of law, the separation of powers, and the recognition of basic civil liberties. It is representative in being based on universal suffrage and periodic relatively free and fair elections of representatives who in turn control government. Democracy in this basic sense offers no guarantees in terms of outcomes. It ensures neither peace, prosperity, nor personal fulfillment. It is a multifaceted and symbolically laden term at the center of a discursive formation that can be put to many, sometimes contradictory, uses. At its most simple, it offers a particular way of trying to structure politics—struggles over some of the defining questions facing us (Who are “we”? How do “we” live together? How do “we” distribute scarce public resources?)—that sees popular government through broad-based and in principle equal participation by all members of the polity as the only legitimate form of government.

Liberal representative democracy is in this sense not only a name for a certain set of political regimes that approximate or lay claim to the above ideals to varying degrees. It is also a truly radical and revolutionary idea. It is today and has throughout history been a notion that challenges the very roots of illegitimate exercise of power both in actually existing democracies that fall short of their self-professed ideals and in regimes beyond them that pay at best lip service to the notion of democracy. The vision offered by liberal representative democracy may seem to be a minimalist vision when compared to various maximalist visions for more deliberative, direct, and/or participatory democracy. But it is a vision nonetheless, and for those inclined to consider such factors, it is a vision that has the relative advantage over maximalist alternatives of being able to offer not only theoretical arguments but also proof of concept from a wide range of different countries around the world, including poor, culturally diverse, and deeply unequal places like India.

I open with these basic observations about democracy because my argument in this essay is that much of the discussion around the relationship between digital information and communication technologies and democracy has focused too little on the question of what connections exist between digital technologies and actually existing, minimalist-vision democracy, and too much on extensive discussion of the possible connections that might be established between digital technologies and alternative, maximalist visions for democracy. In my view, this is a problem in two ways. First, emphasis on alternatives at the expense of the actual limits our understanding of the world we live in. Second, abstract discussions of what digital *might* do for democracy seems to sometimes be allowed to serve as a stand-in for what it *does* do in and for democracy, in turn distorting and exaggerating its role. We need to find a different starting point if we wish to understand the relationship between digital technology and democracy. Actually existing liberal representative democracy and a focus on practice over potential might provide such a point.

Liberal representative democracy is, as Amartya Sen has argued, a "universal value," and historical antecedents, ideas and practices of various forms of popular government, can be found across the world with no direct lineage to "classical," let alone "Western" notions.[2] Democracy is not agreed upon everywhere, let alone practiced everywhere, but it is an idea that is *available* everywhere. It was a powerful part of Asia's intellectual response to Western imperialism around the turn of the century, as it was of struggles for universal suffrage and rule of law in Europe and the United States.[3] In reality, it is never all that we want it to be. Democracy in the full sense of the word will, as Václav Havel put it in his address to a joint session of the US Congress in 1990, "always be no more than an ideal; one may approach it as one would a horizon, in ways that may be better or worse, but it can never be fully attained." Indeed, seen in light of even a minimalist vision of liberal representative democracy as an ideal, polities even approximating democratic regimes are really a twentieth-century phenomenon. It is an ideal whose partial, imperfect, and precarious realization depended upon successful struggles for women's suffrage, for decolonization, and for the recognition of a wide range of civil rights building on the gains of the bourgeois and popular revolutions of the eighteenth and nineteenth centuries. It is an ideal that is never fully realized, as rights are always violated, as some people are always excluded, and as democratic regimes sometimes lapse or even collapse.

Freedom House estimates that 43 percent of the world's population lives in more or less democratic countries by the basic standard provided by liberal and representative democracy. The number increased sharply in the early 1990s after the fall of Soviet communism (also the period in which the self-styled "people's democracies" ceased to effectively contest the term *democracy*), but has changed little in the 2000s and 2010s.[4] The figure includes not only little seemingly picture-book-perfect pockets of peace and prosperity like the Nordic welfare states, but also more complex cases like Brazil, Ghana, India, Serbia, and the United States. In total about three billion people live in such actually existing democracies. They face challenges enough to fill their days, but also enjoy a degree of respite from the combination of domination, exploitation, and arbitrary violence that characterizes much of human political history, and many more around the world have in recent years struggled for this kind of minimal democracy, including people in China, Egypt, Iran, Myanmar, and Ukraine. As much explosive political potential rests in the combination of the Chinese (Mandarin) characters 民 and 主 as in the combination of *demos* and *kratos*, *popular* and *government*, in many other languages.

Digital technologies are increasingly part and parcel both of the everyday political processes of actually existing democracies and of the organizing and mobilization of movements for democratization, just as they are increasingly part and parcel of many other parts of life. In the twenty-five years since the World Wide Web made the internet a more accessible medium for ordinary people, many in high-income countries have moved from occasional access via desktop computers and dial-up modems to constant connectivity via smartphones and mobile broadband access, and digital technologies are increasingly ubiquitous as parts of family life, friendships, work, leisure, consumption—and of politics. More and more people use digital technologies as one of several ways in which they engage with political processes. More and more organized political actors, including political parties, political campaigns, and various sorts of interest groups, have integrated digital technologies in their own operations.[5] On a number of occasions, from the Tea Party and Occupy Wall Street in the United States to the austerity protests across much of Southern Europe and the anticorruption Hazare movement in India, digital technologies have been an important part of how people have organized and mobilized politically to democratize democracies they saw as falling far short of their own ideals and aspirations.[6] Similarly, people organizing and mobilizing for democracy in other contexts have, in countries as different as Iran, Egypt, and Ukraine, relied in part on digital technologies to connect with other activists, coordinate their activities, and publicize their struggle (digital technologies may well have played a more important part in some struggles for democratization than in already-democratic contexts).[7] Spend a day with people actively involved in politics—an elected official, a trade unionist, a movement activist—and you are likely to see them rely on their cell phone, their email, perhaps social media, and almost certainly a whole suite of back-end tools, whether tailor-made software packages, off-the-shelf desktop applications, or cloud-based services, for managing their work. (This practical reliance is of course the motivation for censoring, monitoring, and blocking such tools either permanently or from time to time, just as other means of expression and assembly are restricted.)

It is not clear, however, that the widespread practical use of digital technologies means that they have played the kind of revolutionary role in actually existing democratic and democratizing political practices that has sometimes been forecast. The precise claims made vary, and it seems unnecessary to rehash them here or single out any one futurist. The overall thrust has been clear and broadly shared in a certain genre: digital technologies were expected to facilitate public debate, increase political participation, and challenge incumbent and often hierarchical and exclusionary organizations by making it easier for people to express themselves, take part in things, and organize without formal organizations.[8] Years later, it is not clear that there is empirical support for these hopeful hypotheses. Research till now suggests that digital technologies have so far had (1) modest, (2) mostly internal, and often (3) indirect and institutional implications for democratic and democratizing practices. A few words about each aspect in turn.

First, in terms of the scale and scope of the implications, attempts to assess the effects of digital technology use on political participation have again and again found only modest effects and often a "reinforcement" tendency where digital technology use may correlate with political participation, but mostly in ways where already-engaged groups are even more engaged and less engaged groups are no more engaged.[9] Digital technologies offer easier access more than anything else, but for many, apparently, access is less of a barrier to political participation than inclination (or confidence that even trying is worth one's while).

Second, a number of studies have shown that digital technologies are part of internal (and incremental, gradual, supplementary) changes in how political parties, political campaigns, interest groups, and social movements organize, mobilize, and communicate.[10] In contrast, the track record of digitally first forms of political participation so far seems sporadic and uneven, at least outside the area of tech activism.[11] The movements hailed in recent years by some as "Facebook movements" are at least as much "city square movements"—focused on contesting "meatspace" more than "cyberspace."[12] And while self-expression and attempts at creating an alternative sphere for democratic decision making and political practice have been part and parcel of many of these mobilizations, trying to be the change one seeks is not new and certainly not unique to the digital environment.[13]

Third and finally, digital technologies are integral to much wider decades-old economic and social shifts that are indirectly changing the very position of democracy in the social order in various ways: through the rise of transnational corporations with far-flung supply chains and the globalization of financial markets; through changes in the news industry that, for all its imperfections, help people follow politics from afar; and by enabling and underpinning the maintenance of extralocal communities of cosmopolitan white-collar professionals as much as of diasporas and migrants.[14] (The combination of political deregulation and a globalized and highly unequal economy enabled in part by digital technologies may well change democracy far more, through what seems to be a secular decline of popular government relative to the rise of the administrative state and private corporations, than any number of online opportunities to express oneself, organize, and get involved.) All these large-scale processes at least problematize the traditional (never unproblematic) democratic assumption of a rough overlap of an economy, a state, and a people.[15]

The point here is not that digital technologies mean nothing for democracy, or that the role they play is unimportant or not worthy of consideration. It is simply that the effects are never unambiguously or unproblematically pro-democratic. Serious analysis has left behind the utopian/dystopian dichotomy some time ago in favor of what we, for lack of a better phrase, might call complex realism, whether with a slightly optimistic bent or with a slightly more pessimistic one.[16] We have arrived at the "it's complicated" phase of our understanding of the relationship between digital and democracy. The questions now concern how it is complicated, where, under what conditions, what it means, and for whom. No evidence-based and rigorous analysis that I know of has lent any credence to the more bombastic pronouncements made and sometimes taken by policymakers (and perhaps parts of the public) to represent actual analysis.

Why, then, have digital technologies fallen far short of the democratic visions that have accompanied them (as such visions have accompanied many other new information and communication technologies in the past)? I would suggest that there are three main factors.

First, the relations between digital technologies and democracy have not developed as was imagined in part because much of what was imagined was imagined on the basis of maximalist visions of democracy with considerable theoretical appeal but little basis in reality. This reflects the unhappy coincidence that both contemporary democratic theorizing as practiced in the academy and much of the public speculation offered by the digital avant-garde has been more concerned with alternative, theoretical, maximalist visions of democracy than with actually existing minimalist visions of democracy (or indeed actually existing alternative democratic innovation: experiments like participatory budgeting, etc.). As a wide range of scholars—from radical democrats like Aletta Norval to liberal democrats like Ian Shapiro—have argued, political theory has, with its professionalization as an autonomous academic field and its move away from its roots in "reflexive practice" (think John Stuart Mill versus John Rawls, or Antonio Gramsci versus Slavoj Žižek), become more and more concerned with abstract self-referential argument and less and less interested in analyzing the intellectual and normative dimensions of actually existing democratic practices.[17] A similar flight from reality can be observed among many of the public intellectuals of the digital revolution. Visions of how digital technologies might enable or be part of entirely different kinds of democratic politics are common. Detailed, evidence-based arguments about what digital technologies mean for actually existing democracy are rare. Everyone has a view on the tool of the moment, whether blogs, wikis, or social media. Fewer have bothered to find out what these tools look like and how they work from the point of view of elected representatives, trade unionists, or movement activists. (Some have speculated about what drives the political economy of this kind of tech commentary.)[18] The problem with this development is not the insistence on imagining a better tomorrow. It is that the distance between reality and imagination too often seems to become an excuse for inactivity; that abstract technological potential is too strongly associated with hope, practical reality with disappointment; that the logics driving the production of the most influential ideas can seem parochial and self-interested; and that our visions of digital democracy are too rarely rooted in firsthand experience with actually existing democratic practices, warts and all.

Second, of course, the relations between digital technologies and democracy have not developed as was imagined in part because many other forces have shaped democracy more than has the spread of relatively affordable and accessible digital technologies during the same time period. (Some of these forces in turn themselves intertwined with the development of digital technologies involved in supply chain management, financial speculation, etc.) In many countries, actually existing liberal representative democracy today faces ever-more acute and multifaceted problems of legitimacy, efficiency, institutional integrity, and popular engagement. These problems by and large predate the rise of digital technologies by decades.[19] More and more people in some countries seem disenchanted with democracy in an age of increased economic inequality, social segmentation, political polarization, and perhaps a loss of faith in the very idea of common solutions to common problems (and indeed the notion of a commonwealth)—but this has little to do with digital technologies. (Though we should note that there is no systematic evidence that digital technologies have enhanced democracy more in countries that, like parts of Northern Europe, have not seen these tendencies to the same extent as, for example, Spain and the United States.)[20]

Third and importantly, the relations between digital technologies and democracy have not as was imagined in part because the vast majority of digital technologies were never developed to enhance democracy in the first place. This is a simple and, in a way, obvious point, but one that often seems lost in public debates and social science discussions of the role of digital technologies and their implications for democracy—the internet, for example, is not a "democratic technology" the way the ballot box is a democratic technology, developed specifically to enable popular government. The internet is an infrastructure for sharing digital information developed first and foremost to underpin and enable various government activities (including security-related ones), later personal and professional communication, and increasingly business interests. The democratic effects of the internet, whatever they are, are unintended externalities, not a primary purpose. When we look at what has been invested in the development of digital technologies, digital communication practices, and the infrastructures underpinning them, billions are being spent year in and year out on developing e-commerce, and hundreds of millions are spent on e-government, whereas e-democracy is an afterthought, subject to much talk and a few millions now and then. The Obama campaign may have spent millions in 2008 and 2012 building their own tailor-made digital infrastructure (a small investment compared to what it takes to develop patient record-processing systems for a midsize hospital system). But most political campaigns, electoral or otherwise, have to rely on their own repurposing of off-the-shelf at-hand technologies or standardized franchises offered by a handful of small tech companies and individual consultants specializing in politics.[21] The real digital R&D and investment happens outside politics, outside democracy.

Together, these three factors help explain why digital technologies have not yet delivered on the high democratic hopes that surrounded them. (A fourth factor might of course be time—if the digital revolution is comparable to the print revolution, we are now in the early fifteenth century, the early stages of a sociotechnical transformation that played out over centuries, not simply years or even decades. No matter how imperfect the historical analogy, we certainly have larger and more substantial changes ahead of us than the ones we have already experienced.) They also highlight the importance of looking at the relationship between digital technology and democracy not only through the lens of our imagination and potential, but also through the lens of political practice, of actually existing democracy.

The advantage of understanding the relationship between digital information and communication technologies through the lens of the practical role of digital technologies in actually existing democracies and democratizing practices is that it, in a way, puts digital in its place, as a set of stage props and as part of the setting of the great drama of popular government, not as a main character. Media and politics are mutually constitutive and deeply intertwined, and changes in one (in this case media) invariably influence the other (politics), but not necessarily for the reasons or in the ways often and most prominently suggested, and always in combination with other forces. It also reminds us that democracy is simultaneously enormously disappointing and absolutely amazing—a peculiar outlier in human history—and that the main democratic struggles of our day are not primarily digital. They are primarily democratic. They involve tools of all sorts, analogue and digital, but are driven more than anything else by what Stanley Cavell gestures toward in the quotation used as an epigraph to this essay—the tension between people's visions of perfect democracy (liberal and representative or otherwise), their perception of the present scene of imperfection, and their belief that they can change it for the better.

See in this volume: activism, culture, forum, geek, hacker, internet, personalization, sharing

See in Williams: behavior, bureaucracy, capitalism, collective, consensus, conventional, democracy, determine, elite, expert, ideology, liberal, masses, popular, standards, western

Notes

1 Keane (2009).
2 Sen (1999). As Keane (2009) shows in his history of democracy, it is, contrary to what is sometimes said, not a "western phenomenon."
3 Mishra (2012).

4 See Freedom House's annual "Freedom in the World" report.
5 Bimber (2003).
6 Bennett and Segerberg (2013).
7 Howard and Hussain (2013); Yang (2009).
8 A parallel track of technological pessimism has predicted almost exactly the opposite—that digital media would fragment public debate, encourage people to opt out of political processes, leave us isolated and at home, etc.
9 Schlozman, Verba, and Brady (2010); Boulianne (2009).
10 Karpf (2012); Kreiss (2012); Lievrouw (2011); Nielsen (2012).
11 Coleman (2013); Milan (2013).
12 Pleyers (2014).
13 Melucci (1996).
14 This point was made twenty years ago by Manuel Castells (1995).
15 Sassen (2006).
16 Compare, for example, Chadwick (2013) and Hindman (2009).
17 Norval (2007); Shapiro (2003).
18 Farrell (2013).
19 Pharr and Putnam (2000); Norris (2011).
20 Vaccari (2013).
21 Nielsen (2011).

References

Bennett, W. Lance, and Alexandra Segerberg. 2013. *The Logic of Connective Action: Digital Media and the Personalization of Contentious Politics*. Cambridge: Cambridge University Press.

Bimber, Bruce A. 2003. *Information and American Democracy: Technology in the Evolution of Political Power*. Cambridge: Cambridge University Press.

Boulianne, Shelley. 2009. "Does Internet Use Affect Engagement? A Meta-Analysis of Research." *Political Communication* 26(2): 193–211. doi:10.1080/10584600902854363.

Castells, Manuel. 1995. *The Rise of the Network Society*. Oxford: Blackwell Publishers.

Cavell, Stanley. 1994. "What Is the Emersonian Event? A Comment on Kateb's Emerson." *New Literary History* 25(4): 951–58. doi:10.2307/469384.

Chadwick, Andrew. 2013. *The Hybrid Media System: Politics and Power*. New York: Oxford University Press.

Coleman, Gabriella. 2013. *Coding Freedom: The Ethics and Aesthetics of Hacking*. Princeton, NJ: Princeton University Press.

Farrell, Henry. 2013. "The Tech Intellectuals." *Democracy Journal*. http://www.democracyjournal.org/30/the-tech-intellectuals.php.

Hindman, Matthew. 2008. *The Myth of Digital Democracy*. Princeton, NJ: Princeton University Press.

Howard, Philip N., and Muzammil M. Hussain. 2013. *Democracy's Fourth Wave? Digital Media and the Arab Spring*. Oxford: Oxford University Press.

Karpf, David. 2012. *The MoveOn Effect: The Unexpected Transformation of American Political Advocacy*. New York: Oxford University Press.

Keane, John. 2009. *The Life and Death of Democracy*. New York: W. W. Norton & Co.
Kreiss, Daniel. 2012. *Taking Our Country Back: The Crafting of Networked Politics from Howard Dean to Barack Obama*. New York: Oxford University Press.
Lievrouw, Leah A. 2011. *Alternative and Activist New Media*. Cambridge: Polity Press.
Melucci, Alberto. 1996. *Challenging Codes: Collective Action in the Information Age*. Cambridge: Cambridge University Press.
Milan, Stefania. 2013. *Social Movements and Their Technologies*. New York: Palgrave Macmillan.
Mishra, Pankaj. 2012. *From the Ruins of Empire: The Revolt against the West and the Remaking of Asia.* London: Penguin.
Nielsen, Rasmus Kleis. 2011. "Mundane Internet Tools, Mobilizing Practices, and the Coproduction of Citizenship in Political Campaigns." *New Media & Society* 13 (5): 755–71. doi:10.1177/1461444810380863.
———. 2012. *Ground Wars: Personalized Communication in Political Campaigns*. Princeton, NJ: Princeton University Press.
Norris, Pippa. 2011. *Democratic Deficit: Critical Citizens Revisited*. Cambridge: Cambridge University Press.
Norval, Aletta J. 2007. *Aversive Democracy: Inheritance and Originality in the Democratic Tradition*. Cambridge: Cambridge University Press.
Pharr, Susan J., and Robert D. Putnam, eds. 2000. *Disaffected Democracies: What's Troubling the Trilateral Countries?* Princeton, NJ: Princeton University Press.
Pleyers, Geoffrey. 2014. "From Facebook Movements to City Square Movements." *Open Democracy*, April 3. http://www.opendemocracy.net/geoffrey-pleyers/from-facebook-movements-to-city-square-movements.
Sassen, Saskia. 2006. *Territory, Authority, Rights: From Medieval to Global Assemblages*. Princeton, NJ: Princeton University Press.
Schlozman, Kay Lehman, Sidney Verba, and Henry E. Brady. 2010. "Weapon of the Strong? Participatory Inequality and the Internet." *Perspectives on Politics* 8(02): 487–509. doi:10.1017/S1537592710001210.
Sen, Amartya. 1999. "Democracy as a Universal Value." *Journal of Democracy* 10(3): 3–17. doi:10.1353/jod.1999.0055.
Shapiro, Ian. 2003. *The State of Democratic Theory*. Princeton, NJ: Princeton University Press.
Vaccari, Cristian. 2013. *Digital Politics in Western Democracies: A Comparative Study*. Baltimore: Johns Hopkins University Press.
Yang, Guobin. 2009. *The Power of the Internet in China: Citizen Activism Online*. New York: Columbia University Press.

9

Digital
Benjamin Peters

Every digital device is really an analogical device.[1]
—Norbert Wiener

"The days of the digital watch," the playwright Tom Stoppard once joked, "are numbered." The pun may prove prescient: the keyword *digital*—derived from the Latin *digitalis*, from *digitus* or "finger, toe"—has enjoyed a steady rise from almost nothing before the 1950s to a top-2,500 word in contemporary English that applies to everything from electronics (not only the digital watch, but also the camera, clock, computer, disc, video), to social descriptors (digital divides, natives, and revolutions), to emerging fields of inquiry (digital art, humanities, physics, and studies). Given all this, however, its heyday as a keyword may already have passed: a "digital computer," for example, is almost unheard-of today exactly because they are so common, while its presumed counterpart, "analog computers," are now marked historical oddities. (As the **analog** essay notes, the popularity of the *analog* could arise only after the invention of the *digital*.) Likewise *digital photography* and *digital television* are quickly becoming simply *photography* and *television*. And at the same time, innovations in computing, such as quantum computing, are also moving to disassociate *computing* from *digital*. In other words, the sweeping success of digital techniques has rendered the term a quintessentially twentieth-, not twenty-first-, century keyword. As digital techniques continue to saturate the modern world, we increasingly find the keyword *digital*, understood in its most conventional sense, slouching past its prime.

That conventional sense—in which digital is synonymous with discrete electronic computing techniques—is not nearly deep, broad, or basic enough. The second half of the twentieth century, with its attending explosion of computing industry and culture, obviously stands at the dawn of "digital" discourse, but there remains to be recovered a much deeper and more diverse history of discrete signification techniques.[2] Perhaps the most ancient of the predecessors to digital discourse dates back to the Latin source of the term itself—the original digit, or the index finger. This essay takes that origin point—a digit is an index finger—literally. In it I will explore how digits do what index fingers do—namely, count, point, and manipulate. ("Manipulate" of course is a back-formation from Latin for *handful*—a handful of fingers.) Ever since we evolved extensor digitorum muscles, ours has literally been what media theorist Teil Heilmann calls a "digital condition": digital media do what fingers do.

This is not just to say that we use our fingers to command digital media to execute commands, which is obviously the case. Rather, like fingers, digit media carry out at least three fundamental (Lacanian) categories of actions: digits count the symbolic, they index the real, and, once combined and coordinated, they manipulate the social imaginary. Only the first of these categories is commonplace: the flood of digital devices has made it simple to think of digits as counting and computing discretely numbered objects. But digits do much else too: they also point, index, and reference objects at a distance, as well as combine into new tool suites capable of profound acts of social manipulation, handling, and management. The act of pointing to or indexing nonsymbolic elements of reality is fundamental to signifying systems of all kinds, including (but not limited to) digital ones (see **analog**). Once rendered symbolically interoperable, digits combine computational and referential powers in ways that allow the stewards of digital systems to manipulate elements of that social reality. At the same time, that digital systems point to nondigital elements of reality approximately (without computational precision) helps limit or check the mistaken threats and promises of our current digital age, including the now-dated prophecy of a digital singularity and other forms of technomillennialism.

In short, we foreclose against a fuller understanding of the limits of our digital condition (and what those limits make possible) when we understand digits only computationally. By reviewing how digits have long functioned not only as symbolic counters (computers) but also as real pointers (indexes) and social manipulators, this essay seeks to help deflate, deepen, and rethink what is fundamental about the digital.

Counting the Symbolic: The Triumphs of Digital Computing

A recent publication in *Science* claimed that the total computing power worldwide has enjoyed a staggering compound growth rate of 83 percent every year since 1986.[3] The seeds of this exponential proliferation of digital computing power have been germinating at least since 1946, when the mathematician John von Neumann showed at the first Macy Conference on Cybernetics that all signals can be converted into digital format simply through the introduction of a discrete, symbolic threshold: at or above this level, call the signal one; below that level, call it zero.[4] These artificial thresholds abound in the natural world: the meridian that the sun crosses overhead in the sky is the threshold between morning and afternoon, and the medium of the sundial; by contrast, the typed time of the standardized clock suppresses and supplants the real time of the sun overhead.[5] While the history of discrete computing traces back at least to Leibniz, the history of discrete or digital computing took wing after the wartime invention and subsequent industrialization of information science by the academic-military-industrial complex on both sides of World War II.

The point to information science, first articulated by Leibniz and later formalized by the logicians Boole and Shannon, is simple: all real signals can be reduced, with certain loss, into digital symbols. Anything one wants to describe—say, content (sensory experience), space (coordinates), time (intervals), or instructions (programming, algorithms)—can be expressed in the irreducibly countable alphabet of that onc binary difference, 0 or 1. As the logician Alan Turing showed in 1937, the most basic digital computer, given enough time and memory, can solve any computable problem.[6] Since then "universal Turing machines," or general-purpose digital computers, have pioneered the spread of generative digital devices.[7] Supported not only by a global computing industry but by a global computerized economy, modern mediated life now proceeds at the pace of networked computing techniques that render all things countable.

The momentous logic of digital computing, taken to its extreme, leads to the position in vogue among digital theorists that everything that is, is in fact countable. Information physicists, for example, contend that nature has always already been digital, or the real is at base symbolic: magnets have north and south poles, electrons are positively or negatively charged, and quarks spin either up or down. Matter itself appears to follow discrete logics of off and on, 0 and 1. In media theorist Friedrich Kittler's phrase, only that which is switchable can be ("nur was schaltbar ist, ist überhaupt"), or—as the theoretical physicist John Wheeler rephrased the atomist worldview that information is not just what we learn—it is what we *are*: "it from bit."[8]

It is as if in the beginning was the bit, and the computing of bits—from stone coins to Bitcoin—has since become the currency of modern life. The effects of precise computation and copying abound. Writing and programming (from glyphs to ASCII code) reduce thought to the graphism of uniquely encoded symbols; so too does cognitivism seek to distill the vagaries of memory, emotion, and experience into the biomechanics of synaptic firings across neurological circuits. In his monumental *The Culture of the Copy*, Hillel Schwartz has claimed that the defining characteristic of modernity (one fully embodied in the digital age) is its preoccupation with exact copying and its discontents.[9] This then is the first feature of the long legacy of the digital: metadata aside, digits are copied with uncanny exactness. It is hard to overlook the ascendance of this one—but only one—kind of fundamental work our digits do: counting, at scales so large and steps so sophisticated that we name the qualitative change in counting *computing*.

The more digital media spread, the more exacting and consuming our counting regimes appear to become. As early as 1936, critic Walter Benjamin pointed out that the mechanical reproducibility—or computational copyability—of content brings with it a new aesthetics: the work of art since modernism and the interwar period has become increasingly imitable and popular, foreshadowing contemporary remix, DIY, pastiche, and bricoleur cultures online and off.[10] Our enthusiasms for the spread of digital counting continue: big data are said to scale computing power from sample set to the whole population of data. (Why consult a book when algorithms scour the whole of the Library of Congress?) Democracy enthusiasts too extol the virtues of online voting and debate, where all voices might count equally (see also **activism**, **democracy**). Chess enthusiasts hunger after—and fear—a complete book of moves online. The clean background of the Google search page obscures a messy algorithm that tempts us to imagine that Borges's all-containing catalog of catalogs lies in reach, just a few finger strokes away.

This digital Matthew effect, where the digital gets more digital, meets its culmination in the simultaneous dream of the information theorist, the universal strategist, the advertising executive, and the utopian futurist: the coming digital "singularity," a term coined by Stanislaw Ulam in 1958 suggesting, in light of von Neumann's discoveries, a coming technologically driven paradigm shift in the history of the human race.[11] Since the most fundamental building block of all that we know and are is already the bit of information—these computation enthusiasts contend—the broader the spread of digital media, the more powerfully certain humans will be able to represent and reshape reality itself. In fact digital computation renders more and more of the world visible to those with the tools to compute, index, and manipulate data. Viewed from the perspective of those occupying the commanding heights of computing alone, digital computing promises nothing less than a total convergence, a singular universe in which all bits are known and in play at once—a worldview the rest of this essay seeks to limit.

Indexing the Real: How Digits Point Elsewhere

Digits certainly compute, but they also do far more than that. Like fingers, they also *point*. And, as anyone who has been burned by a misplaced finger knows, pointing is far from an exact science. Just as the internal systems digital media compute are finite, rational, and discrete, so too must the external world to which the same media point remain infinite, irrational, and approximate, and it is this difference that firmly insures against both the promise and the threat of total digital convergence.

Consider the longer view that emerges once we see digital media as those media that, like our fingers, count the symbolic, point to the real, and manipulate the social imaginary. In this light, digital media include the finger as the original extension of the human body, the coin, the *yad* ("hand" or Torah pointer), the manicule (or "pointing hand," "index," or "digit" in the margins of eleventh- to eighteenth-century typography), the piano keyboard, filing systems, the typewriter, and the electronic telegraph. All these media, among many others, are digital in the simple sense that humans interface with them *digitally*, or with our fingers via manual manipulation and push buttons. Fingers and digital media alike flip, handle, leave prints, press, scan, sign, type. The touchscreens we pet and caress today continue the age-old work of counting, pointing out, and manipulating the literate lines animating every modern media age, including our own. Digital media, such as these, point and refer to real-world objects outside of themselves, and this transducing from the symbolic to the real limits both the computing *and* the indexing power of digital media.

Another name for a digit that points is an index (or its plural, *indexes*). Charles Sanders Peirce, a founding pragmatist and semiotician, divided the world into three types of signs (unlike the Saussurean signifier-signified binary behind the postmodern turn): the icon, which like a portrait resembles the thing it points to; the symbol, which, like the word *couch*, means a place to sit only because convention has taught us to recognize the arbitrary name as meaningful (or as Shakespeare put it, "a rose by any other name would smell as sweet"); and the index, which has a natural connection to the thing it points to but it *not* that thing itself, such as how a symptom points to a disease while not being the disease, or an anthill points to ants without resembling ants. Much work has been and could be done considering how coins, manicules, and files precede digital media in pointing to, without resembling, the semiotic regimes that organize life, such as economic currency (e.g., the head of a leader on a coin), text (e.g., the hand that learns to read by skimming along the line), or bureaucracies (e.g., the metadata markings on a file that place that file in a larger set of procedural operations).[12] Just this hint at the various ways digital media index the world beyond numbers helps upend a narrow-minded focus on computing as a harbinger of digital convergence. Digital media have long indexed the world.

To be an index is to render approximately or refer to something outside of its own signifying system, and thereby to claim some nonnecessary but useful connection to that thing. Indexes abound: a book index points the reader from outside the body of a text to the right page in the body of the text, but not the exact phrase. The page number in an index reference is not the quotation itself, but it helps approximate its location. The weather vane is not the wind, but it indexes that complex vector field of air into a single well-defined direction. Likewise smoke indexes fire: smoke is not fire, but it signals fire by saying, roughly, "Follow me to an ongoing combustion process." For philosophers of language from Wittgenstein to Austin, this point is basic: all meaningful relationships begin by creating a semiotic structure that excludes something else. (This is true in a romantic sense as well.) For signifying systems of all kinds, the structure of meaning is indexical.

Digital media thus have meaning insofar as they index the world. They point beyond themselves and exclude something significant in the process. Indexical exclusion holds computationally, as Gödel's famous theorems prove that no computational system can be both complete and consistent on its own terms.[13] It is also true socially: our favorite social networking sites reacquaint us with friendly personas and profiles that point to but are not the friends we know in person. Google Maps gives a godlike view of the land surface we both know and do not know by presenting a scalable approximation of it, but it does not give us the land itself. (To represent reality exactly, a map would cost in computing power at least as much as the reality it indexes.)[14] Digitally programmed artificial intelligence, robots, prostheses, 3D printers, and animation serve modern humans because they imperfectly imitate other natural objects. No media copy natural objects exactly: clones, duplicators, alternate realities, perfect memory—this is the stuff of Silicon Valley hype and science fiction dystopia. That original digit, a human index finger, is useful exactly *because* a finger is not the object it refers to. In other words, it is precisely the negotiable yet natural relationship of all indexes to their referents that makes digital media do more than render the world computationally. Digital media also render the world open and inexactly with a flow of perpetual references elsewhere: that digits (think fingers) can point elsewhere is what grants them their fundamentally analogic character—it is what gives digital work, like all work, the possibility of meaning. As Norbert Wiener remarked, "Every digital device is really an analogical device," although only in a narrow sense (see **analog**).[15]

It is perhaps fitting then that Claude Shannon, Wiener's contemporary and colleague, launched information theory by excluding meaning itself. In his landmark 1948 article that ushered in a strictly computational approach to communication championed in the first third of this essay, Shannon began with this striking constraint: no act of computing (or counting) alone can claim to understand how its messages relate to the real world. *Five apples* means something in the real world, but *five* by itself does not. For him, computing and indexing are functions as distinct as fingers that count and fingers that point. He describes the indexing function thus: "Frequently the messages have meaning; that is they refer to or are correlated according to some system with certain physical or conceptual entities." Then he adds famously, "These semantic aspects of communication are irrelevant to the engineering problem."[16] He is not saying that digital media do not shape our world; rather he is sagely acknowledging that a computational understanding of digits can never speak to such matters. In other words, Shannon, a founder of modern computing, begins by effectively dismissing the premise behind any promise of a digital singularity or computational convergence. He is not saying that digits do not have meaning in the world; he is saying only that the question of meaning is irrelevant so long as we understand digits as only those things that count.

Precise computing and inexact indexing coexist quite happily. Consider probability, the mathematical engine of information theory. Probability is clearly computational, and yet it continuously and uneasily indexes a world that cannot be totally counted. In other words, the modern probabilistic relationship to reality is foremost indexical, before it is even symbolic. Probabilities do not just count what is. They point ahead, with certain uncertainty, to what could be—to the future or elsewhere. To say, for example, that there's a 42 percent chance of rain tomorrow sounds mundane, but it actually exercises an extraordinarily imaginative license to infer from past data about multiple distinguishable futures—or, given a hundred future tomorrows, forty-two will experience rain. (Perhaps, like the weather, nothing digital about the future is singular or certain.)

In order to be computable, all digital messages must first be treated as if they were part of a possibility index—or, as Shannon puts it, "one selected from a set of possible messages."[17] Even though many messages we send and receive likely have some meaning, the vast majority of mathematically possible messages are sheer spam (Borges's Library of Babel again makes this point). By understanding digits as indexes we return to a familiar point: a finger, like digital media, can point to anything, which means that what we point to is probably meaningless—and at best probabilistically meaningful—most of the time.

We can now see how the digital and the analog are nonoppositional modes of indexing the world. Take the classic analog medium, the phonograph (an early record player named for how it transduces a real-world event of sound, *phono*, into symbolic writing, *-graphy*, and then reads the writing into reproducing the sound). Phonography transforms continuous grooves on a record into continuous sound waves in the air. Both analog and digital techniques, in other words, index the real approximately—and they do so differently and they do so nonexclusively. (There are many other kinds of media.) To imagine the opposite—that digital and analog are in exclusive opposition—is unthinkable: first, imagine that digital media and analog media were in fact the only ways of representing the world. Now suppose somehow there were a total convergence between digital computing and the real world (imagine your smartphone contained the whole world precisely). Even were this to somehow be the case, the end relationship between the digital profiles of the contacts in your phone and the real-world people you know, or between the symbolic and the real, would be—as it always has been—indexical, not computational.[18]

Manipulating the Social: The Discontents of Digital Power

Digits, like fingers, can wag, curl, clench, and deliver crushing blows. The spread of digital computing is no unmitigated good for all, and especially for the disenfranchised many. As Langdon Winner predicted decades ago, the larger the franchise, the more computing power stands ready to serve its interests.[19] Whatever else big data may mean, big data surely means big data brokers. Digital media index not only our world but all known possible worlds—and this means, in practice, those parts of the world that many would prefer not to consider. The cascading collapse of privacy is not only a sweeping narrative of decline—it demotes the status of the modern individual to one on par with most humans in history: we are all again exposed to the elements and subject to powers far greater than ourselves.

This fact sobers digital convergence hype and at once highlights both the true negatives and the false positives behind consequential social problems brought about by digital media. Anyone tempted to believe in the coming computational convergence need only observe how rarely online avatars and dating profiles resemble their users. Symptoms of you and me lurk online. Digital media and real-world actors do not index each other perfectly: they manipulate one another in both directions, although still unevenly. It would not be ridiculous for a Facebook user, for example, to not "friend" other users they have not met in real life. (Social meaning manipulates what exactly is social about social networks.) However, it would be a form of madness to run the relationship between social network and real world in the other direction: strange would be the sociopath who elects to stop being friends with people in real life because their comments are not on their Facebook feed. (How far more frequently do we stop feeling friendly toward them because their comments are!) In other words, our digital registers need not resemble our real-life registers, and vice versa: indexes point in one direction at a time. And yet, of course, Facebook is no dormant register: it is an active institution algorithmically manipulating our social experience (see **algorithm**). These and other social network platforms filter "friends" and "followers" from our view all the time: digits, combined into corporate platforms, manipulate and promote, fix unseen connections, collapse our many social selves into one persona, privatize our privacy, and flood and flush the marketplace of attention with its wares. It has long been obvious that humans use media to handle modern real life. It remains less obvious how the powers behind digital media handle us.

Consider again how Google Maps, a modest indexing compared to the digital ears and eyes of surveillance states, represents not only the relevant roadsides digital users seek. It also indexes and puts on display images of the homeless, those accused without trial, and all others whose presence and privacy our law, technology, and society do not defend.[20] By indexing all that we send, receive, and process into distant databases, cloud computing techniques force users to exist in a world that can only be "saved," and rarely deleted, with a click of a finger (see **cloud**). Digital databases sometimes index the social with eerie accuracy: a recent study found, for example, that the metadata alone collected in NSA phone tapping have enough inferential power to invade personal privacy.[21] The same indexes risk condemning us with errors, both our own and its own: rumor holds that purchasing a Union Jack and certain soil fertilizers may be enough to automatically place an American citizen on a national terrorist watch list that had swollen by 2013 to include over 875,000 individuals, arguably almost all of them false positives.[22] Digital techniques let those in privileged positions symbolically construct models of the world that index and manipulate it. Digital hands take many shapes: at times the hands of the large and unscrupulous data manipulators take care to caress and palliate those who serve them; at other times the subtle hands of the big data manipulators surgically excise bits from our digital personas and body politic; at others the hands at work paper over and screen from our view the alarming costs of mounting ecological and other social crises; at still other times the hand, curled into the fist of social rage or the martial strikes of cyberwarfare, hovers precariously in the air, threatening to crush its choice target.

In short, digits have never been just computing symbols. Digital techniques—tools ever *in* and *of* our hands—both index the real world and manipulate our many social worlds. Not only has it been obvious since Shannon that digital convergence is a priori impossible; more important, the necessarily imperfect indexes that confound the relationship between the symbolic and the real, or between what counts and what is, compel us to recognize profound social problems that attend the increasingly rapid, uneven, and worrying concentration of computational power and resources.

Conclusion

Digital media have been counting the symbolic, pointing toward the real, and manipulating lives since humanoids have had fingers, even though the explosion of computing power has swept up the digital to such a degree that the techniques may now be outrunning the term. We can now count down the numbered days of the keyword *digital*, to rehearse Stoppard's jest. To understand our digital age, we must understand not only the numbers—that digits count, compute, construct, and copy internally discrete symbolic worlds—but that digital media can point to or index all possible worlds, not only our real one. This second point helps counterweight, sober, and caution Whiggish enthusiasm for the ongoing digital revolution leading to total media convergence or a technologically determined single future.

The work of digital media can be said to rest at our fingertips. The work of digital computing is similar to counting on our fingers: we think counting is abstract and without obvious real-world unit, and yet counting takes place on the very handy extensions of ourselves—digits, media, and their combination—that permit our bodies to interact with and to manipulate a material world. The human species has always already been born digital: building tools that count, index, and manipulate the world is almost unique to the anthropoid species—those higher primates with digital tools built right into their hands. While counting 1 + 1 = 2 on our fingers is computationally exact, to do so is to engage in higher abstraction: without a unit or referent, the number "2" remains a quantity without qualities in the real world. Only by indexing our counting to real-world objects do we embody our computational abstractions. Human hands, in other words, are the first digital medium to don real-world units that apply with probabilistic, and never precise, degrees to all possible worlds around us. By pointing or orienting ourselves to different objects, our digits have long manipulated the world around us. This is nothing new: what is new is the commanding degree and scale to which, in the past seventy years or so, trivially large reservoirs of computing power have begun to be consolidated in the hands of increasingly powerful data-rich institutions—corporation and state alike—and much less so self-organizing groups of people. Socioeconomic privilege continues to scale with digital privileges. (The belief in sousveillance as a viable way of resisting institutional surveillance is most concentrated among affluent technoactivists.)

These trends suggest that the consequences of computational power itself will not converge, and there is no reason to imagine that the institutions that command its powers will (want them to) either. Rather the far more awesome power resting in the hands of our digital species is to point to and manipulate any number of modeled worlds. There remains much to be done to model more equitable and sustainable worlds. Perhaps we can begin by understanding the digit as an openly imitable and probabilistically imperfect index of any thinkable world, including this world, with which there can be no final convergence. The last seventy years have ushered into existence a host of digital devices that now populate our pockets, warehouses, and working models of the world. The lot of these reality doppelgängers, like that of all indexical media before them, is to point to endless and imprecise imitations of their makers.

See in this volume: algorithm, analog, cloud, culture, democracy, event, information, mirror, personalization, sharing

See in Williams: bureaucracy, capitalism, ideology, jargon, mechanical, rational, representative, technology

Notes

1 Quoted in Claus Pias, (ed., *Cybernetics/Kybernetik. The Macy-Conferences 1946–1953*, 2 vols. (Berlin: Diaphanes, 2003).

2 For more on that long history, see Bernard Siegert, *Passage des Digitalen: Zeichenpraktiken der neuzeitlichen Wissenschaften 1500–1900* (Berlin: Brinkmann & Bose, 2003).

3 Martin Hilbert and Priscila Lopez, "The World's Technological Capacity to Store, Transmit, and Compute Information," *Science* 332(6025) (2011): 60–65.
4 John von Neumann gave the first talk delineating analog and digital at the 1946 Macy Conference, and elaborates on how, for example, organisms might be treated as digital phenomena in John Von Neumann, "The General and Logical Theory of Automata," in *Cerebral Mechanisms in Behavior*, ed. L. A. Jeffress, The Hixon Symposium (New York: Wiley, 1951). See also Steven J. Heims, *The Cybernetics Group* (Cambridge, MA: MIT Press, 1991).
5 The psychoanalyst Lacan might say that analog-to-digital conversion seeks to suppress "the real" with "the symbolic," and here I understand, with media theorist Kittler, "the real" as those physical, continuous, material, and analog elements of our world that can be recorded by a phonograph, while "the symbolic" makes up all the artificial, discrete, logical, and digital elements that can be recorded by a typewriter. Cf. Friedrich Kittler's *Phonograph, Film, Typewriter* (Stanford, CA: Stanford University Press, 1999).
6 Alan Turing, "On Computable Numbers, with an Application to the Entscheidungsproblem," *Proceedings of the London Mathematical Society*, ser. 2, 42 (1936–37): 230–65.
7 Jonathan Zittrain, *The Future of the Internet, and How to Stop It* (New Haven, CT: Yale University Press, 2008).
8 Friedrich Kittler, *Draculas Vermächtnis. Technische Schriften*. (Leipzig: Reclam, 1993/2003), 182; John Archibald Wheeler, "Information, Physics, Quantum: The Search for Links" in *Complexity, Entropy, and the Physics of Information*, ed. W. Zurek (Redwood City, CA: Addison-Wesley, 1990).
9 Hillel Schwartz, *The Culture of the Copy: Striking Likenesses, Unreasonable Facsimiles* (Cambridge, MA: MIT Press, 1996). \
10 Walter Benjamin, "The Work of Art in the Age of Mechanical Reproduction," in *Illuminations*, ed. Hannah Arendt (London: Fontana, 1968), 214–18.
11 S. Ulam, "Tribute to John von Neumann," *Bulletin of the American Mathematical Society* 64 (1958): 1–49.
12 For more on C. S. Peirce, see T. L. Short's *Peirce's Theory of Signs* (Cambridge: Cambridge University Press, 2007).
13 For more, see Kurt Gödel, "On Formally Undecidable Propositions of *Principia Mathematica* and Related Systems I," in *Kurt Gödel Collected Works*, ed. Solomon Feferman (Oxford: Oxford University Press, 1986), 1:144–95.
14 John Durham Peters, "Resemblance Made Absolutely Exact: Borges and Royce on Maps and Media," *Variaciones Borges* 25 (2008): 1–23.
15 Quoted in Pias, *Cybernetics/Kybernetik*. See also epigraph.
16 Claude Shannon, "A Mathematical Theory of Communication," *Bell System Technical Journal* 27(3): (1948): 379–423, see 379.
17 Ibid., 379.
18 See Jonathan Sterne, *MP3: The Meaning of a Format* (Durham, NC: Duke University Press, 2012). Also Sterne, "The Death and Life of Digital Audio," *Interdisciplinary Science Reviews* 31(4) (December 2006): 338–48. Eric

Rothenbuhler and John Durham Peters, "Defining Phonography: An Experiment in Theory," *Musical Quarterly* 81(2) (Summer 1997): 242–64.

19 Langdon Winner, "Mythinformation," in *The Whale and the Reactor: A Search for the Limits in an Age of High Technology* (Chicago: University of Chicago Press, 1986), 3–18.

20 Siva Vaidhyanathan, *The Googlization of Everything (and Why We Should Worry)* (Berkeley: University of California Press, 2011), 106–7.

21 Jonathan Mayer and Patrick Mutchler, "Metaphone: The Sensitivity of Telephone Metadata," http://webpolicy.org/2014/03/12/metaphone-the-sensitivity-of-telephone-metadata/.

22 See the "Information Sharing Environment Annual Report to the Congress 2014," https://www.ise.gov/resources/document-library/ise-annual-report-congress-2014.

References

Benjamin, Walter. "The Work of Art in the Age of Mechanical Reproduction." In *Illuminations*, edited by Hannah Arendt, 214–18. London: Fontana, 1968.

Gödel, Kurt. "On Formally Undecidable Propositions of *Principia Mathematica* and Related Systems I." In *Kurt Gödel Collected Works*, edited by Solomon Feferman, 1:144–95. Oxford: Oxford University Press, 1986.

Heims, Steven J. *The Cybernetics Group*. Cambridge, MA: MIT Press, 1991.

Hilbert, Martin, and Priscila Lopez. "The World's Technological Capacity to Store, Transmit, and Compute Information." *Science* 332(6024) (2011): 60–65.

Kittler, Friedrich. *Phonograph, Film, Typewriter*. Stanford: Stanford University Press, 1999.

———. *Draculas Vermächtnis*. Technische Schriften. Leipzig: Reclam, 1993/2003.

Mayer, Jonathan, and Patrick Mutchler. "Metaphone: The Sensitivity of Telephone Metadata." http://webpolicy.org/2014/03/12/metaphone-the-sensitivity-of-telephone-metadata/.

Neumann, John von. "The General and Logical Theory of Automata." In *Cerebral Mechanisms in Behavior*, ed. L. A. Jeffress, 1–31. The Hixon Symposium. New York: Wiley, 1951.

Peters, John Durham. "Resemblance Made Absolutely Exact: Borges and Royce on Maps and Media." *Variaciones Borges* 25 (2008): 1–23.

Pias, Claus, ed. *Cybernetics/Kybernetik. The Macy-Conferences 1946–1953*. 2 vols. Berlin: Diaphanes, 2003.

Rothenbuhler, Eric, and John Durham Peters. "Defining Phonography: An Experiment in Theory." *Musical Quarterly* 81(2) (Summer 1997): 242–64.

Schwartz, Hillel. *The Culture of the Copy: Striking Likenesses, Unreasonable Facsimiles*. Cambridge, MA: MIT Press, 1996.

Shannon, Claude. "A Mathematical Theory of Communication." *Bell System Technical Journal* 27(3) (1948): 379–423.

Short, T. L. *Peirce's Theory of Signs*. Cambridge: Cambridge University Press, 2007.

Siegert, Bernard. *Passage des Digitalen: Zeichenpraktiken der neuzeitlichen Wissenschaften 1500–1900*. Berlin: Brinkmann & Bose, 2003.

Sterne, Jonathan. "The Death and Life of Digital Audio." *Interdisciplinary Science Reviews* 31(4) (December 2006): 338–48.

———. *MP3: The Meaning of a Format*. Durham, NC: Duke University Press, 2012.

Turing, Alan. "On Computable Numbers, with an Application to the Entscheidungsproblem." *Proceedings of the London Mathematical Society*, ser. 2, 42 (1936–37): 230–65.

Ulam, Stanislaw. "Tribute to John von Neumann." *Bulletin of the American Mathematical Society* 64 (1958): 1–49.

Vaidhyanathan, Siva. *The Googlization of Everything (and Why We Should Worry)*. Berkeley: University of California Press, 2011.

Wheeler, John Archibald. "Information, Physics, Quantum: The Search for Links." In *Complexity, Entropy, and the Physics of Information*, ed. W. Zurek, 3–28. Redwood City, CA: Addison-Wesley, 1990.

Winner, Langdon. "Mythinformation." In *The Whale and the Reactor: A Search for the Limits in an Age of High Technology*, 3–18. Chicago: University of Chicago Press, 1986.

Zittrain, Jonathan. *The Future of the Internet, and How to Stop It*. New Haven, CT: Yale University Press, 2008.

10

Event

Julia Sonnevend

An event—to borrow and rephrase a popular line—"is what happens to you while you're busy making other plans." Some events happen because they are planned—for instance, weddings and presidential inaugurations—while other events are sudden shocks, like cancer diagnoses and assassinations. Certain events gain significance beyond a family or a community and become public events. Events can even turn into what I call "global iconic events" (Sonnevend 2013b, 2015) that international media cover extensively and remember ritually. These events quickly occupy our old and new media, triggering the peaks and troughs of social media trend lines. Planned or unplanned, minor or earth-shattering, digital or analogue, all these events do the same thing: they structure our social lives and give reference points for our life stories and global histories.

The Oxford English Dictionary defines *event* as "a thing that happens or takes place, especially one of importance." Its English-language history dates back to the sixteenth century. *Event* originates from the Latin *eventus*: from *evenire* "result, happen," from *e* "out of" and *venire* "come." Both the definition and the etymology of *event* indicate an influential and dynamic phenomenon: events are *important happenings*. Interestingly, while events are crucial for our mediated social lives, a comprehensive theoretical concept of events has not emerged within media research. It is easy to understand why media scholars relegated events to philosophers, historians, and sociologists. Events misbehave and innovate as often as teenagers do. Some events are idiosyncratic, contourless, and quite resistant to categorizing, while still others occur so frequently they escape attention altogether.

Some media researchers have nonetheless wrestled with events, especially with those historic occasions that "shook the world." For instance, Amit Pinchevski and Tamar Liebes (2010) wrote about the media coverage of the Eichmann trial, Daniel Hallin (1986) and Marita Sturken (1997) analyzed the media constructions of the Vietnam War, and Barbie Zelizer (1992) examined the media representations and retellings of the Kennedy assassination. Some scholars have moved beyond the particular case study analysis to define whole genres of media events such as media scandals (Lull and Hinerman 1997), disaster marathons (Liebes 1998), media spectacles (Kellner 2003), public apologies (Kampf 2009), and rituals of excommunication (Carey 1998). The most comprehensive and creative analysis of media events came from Daniel Dayan and Elihu Katz, *Media Events: The Live Broadcasting of History* (1992), whose work focused on preplanned, celebratory events covered *live* by television. Mapping onto Max Weber's treatment of rational-legal, charismatic, and traditional sources of authority, Dayan and Katz presented three scripts of media events. These were *contests* (for instance, the Olympic Games and the Watergate hearings), *conquests* (such as the landing on the moon and Pope John Paul II's visit to Communist Poland), and *coronations* (for example, the funeral of President Kennedy and the royal wedding of Prince Charles). Many scholars have subsequently built on and critiqued Dayan and Katz's concept of media events (Scannell 1996, 2014; Schudson 1992; Couldry, Hepp, and Krotz 2010).

A broader discussion of "events in media," nonetheless, must move beyond the somewhat confined scope of Dayan and Katz's canonic treatment of "media events." What about events that do not have *live* coverage (like the Cambodian genocide), events that are not covered by television (like the Eichmann trial in Israel), and events that are celebrated in one country but not in another (the fall of the Berlin Wall in American and Soviet media)? In other words, what about events that are covered by media but not in the particular genre of what Dayan and Katz called "media events"? And, more specifically, how could we theorize events in a "digital age" when stories can quickly spread globally—often leaving too much or too little trace behind?

Here I will consider "events in media," including but not limited to the narrow genre of "media events." I will analyze events with a framework that accounts for both digital and predigital events. Since events predate the "digital era," and will certainly outlast it, their analysis can be done only in a framework that looks at the digital age as merely *one period in history*—a period, nonetheless, that makes certain aspects of events in media more prevalent or salient. Legal scholar Jack M. Balkin has persuasively argued for this methodological approach:

> In studying the Internet, to ask "What is genuinely new here?" is to ask the wrong question. . . . Instead of focusing on novelty, we should focus on salience. What elements of the social world does a new technology make particularly salient that went relatively unnoticed before? What features of human activity or of the human condition does a technological change foreground, emphasize, or problematize? And what are the consequences for human freedom of making this aspect more important, more pervasive, or more central than it was before? (Balkin 2004, 53)

I will take up four features of "events in media," highlighting how the digital era makes each feature more salient: (1) the power of the occurrence vis-à-vis its narrative as an "event," (2) the witnesses who tell the story of an "event," (3) the embodiments of the "event" in a variety of media, and (4) the travel of "events" across cultural and geographic boundaries.

Occurrences and Events

Every event consists of some *occurrence* on the ground and a related *narrative* of an event. The systematic mass murder perpetrated during World War II, originally narrated as a series of "atrocities," became a moral universal in the West, described over time as the "Holocaust" (Alexander 2002). After four planes were deliberately crashed in the United States on September 11, 2001, these occurrences together received the name "9/11." A sequence of occurrences in a small Connecticut town (the shooting in the Sandy Hook Elementary School in Newtown, President Obama's visit to the town, the funerals of the victims, the investigation into the shooting, protests for new gun laws, etc.) were all summarized under the Twitter hashtag "SandyHook." In all these cases a myriad of *occurrences* were pulled together and interpreted in a *narrative* of a certain genre of an "event."

While narratives and naming practices appear to be powerful tools in shaping events, they are not omnipotent. Consider the example of terrorist attacks. These occurrences can be narrated in opposing ways, as acts of wanton destruction or as acts in observance of a higher moral order. A good example of framing a terrorist attack as a regrettable but unavoidable "must" appears on a plaque on the King David Hotel in Jerusalem: "The hotel housed the Mandate Secretariat as well as the Army Headquarters. On July 22, 1946, [Zionist paramilitary] Irgun fighters at the order of the Hebrew Resistance Movement planted explosives in the basement. Warning phone calls had been made urging the hotel's occupants to leave immediately. For reasons known only to the British, the hotel was not evacuated and after 25 minutes the bombs exploded, and to the Irgun's regret and dismay, 91 persons were killed." This original wording infuriated the British insofar as it suggested that the British, not the Irgun, were responsible for the attack. Although the wording was subsequently revised, the final sentence, including the phrase "regret and dismay," remained.

This excerpt shows the power of narratives in shaping occurrences into certain types of events, but it does *not* prove that narratives can do anything. We can narrate a terrorist attack as a crime or as an accident, but hardly as a wedding. In other words, there are multiple, but *limited*, ways to read events. This feature of events is even more salient in the digital age: while there are seemingly limitless ways to express ourselves on diverse platforms, occurrences do still shape our narration of events. The abundance of digital speech platforms makes the limitations of our narratives more visible: no matter how much and how quickly we speak, a terrorist attack cannot be told as a wedding in our narratives. Occurrences still set boundaries for digital narratives.

Witnessing an Event

Who sees and tells the story of an event, who writes its "birth certificate," is central to every event's existence. Storytellers are required to bind occurrences together and elevate them into an "event." In other words, events need witnesses (Peters 2001). Media witnessing occurs in three distinct forms: witnesses *in* media (when witnesses of the occurrence share their experiences in media), witnessing *by* media (when journalists bear witness to occurrences), and witnessing *through* media (when audiences are positioned by media as witnesses to occurrences—for instance, when people watch live coverage of events on television) (Frosh and Pinchevski 2009).

These various forms of witnessing all shape the boundaries of events and communicate them to distinct primary and secondary audiences. In digital environments, the first category gains prominence: witnesses of events rapidly share their experiences on social media, often providing the initial framings of the event. If journalism is the first rough draft of history, as the phrase goes, perhaps social media accumulate the notes behind it. All events have competing witnesses who actively spread their narratives, and, as a result, there are immediately diverse sets of contrasting stories. The battle of a variety of instantaneous digital narratives and counternarratives leads to intense discussions—and often to quick forgetting.

Embodying Events in a Variety of Media

Events are more vulnerable than we would think. We easily forget them. We do this not only with birthdays and anniversaries, but also with major historic events. Each generation has its own events that it regards as earth-shattering. For instance, certain generations have "flashbulb memories" of the atomic bombing of Hiroshima, while other generations remember keenly the moment they received the news about the attacked Twin Towers, the death of Michael Jackson, or the inauguration of the first African American president. But an iconic event to one generation often appears mundane to the next. Events are heavy: it is hard to carry them across time, space, and media.

Therefore, in order to endure in recitations that cross generations, occurrences need more than memorable narratives that construct them as mythical, resonant "events." They also need to be carried by a diversity of media. Even the seemingly most powerful and visually spectacular event cannot survive the passage of time without substantial narrative presence across multiple media. For instance, consider all the commemoration efforts undertaken to keep the memory of 9/11 alive: the names of victims are read aloud at Ground Zero at every anniversary, a huge cosmopolitan museum has recently opened in New York, and the event's story is embodied in countless social media campaigns, souvenirs, documentaries, and history books. Selfies taken at the 9/11 memorial occupy Instagram, Facebook, Flickr, Twitter and many other social networking sites. Owing to the event's significance and its current omnipresence, those who personally remember the day of September 11, 2001, may deem it unforgettable, even though it is not. Few college freshmen today have acute personal memories of the event that took place over a decade ago; its lasting resonance will require continued promotion of the event's simple narrative and spectacular imagery across "old" and "new" media alike. And there are many less "spectacular" events that capture the imagination of media users. Will we discuss Ferguson, Obama's immigration reform, or the recent Ebola outbreak a year—or a decade—from now? Nevertheless, these events occupy my Twitter feed (and perhaps yours) at the time of writing (November 2014). They resonate with large audiences and are indexed by popular Twitter hashtags at the same tremendous speed that they, in turn, disappear from public consciousness.

Beyond Boundaries: Events That Travel

Most events are narrated locally—on the national, regional, or social group level. But some events receive powerful and lasting transnational narration. A transnational narrative needs to be simplified and universalized; it has to remove the event's original complexity and context, thus making it transportable across boundaries. There are at least five dimensions of a global iconic event's narration: (1) foundation: the event's narrative prerequisites; (2) mythologization: the development of the event's elevated language and lasting message; (3) condensation: the event's encapsulation in a brand—a simple phrase, a short narrative, and a recognizable visual scene; (4) counternarration: competing stories about the event; and (5) remediation, when the event's brand travels across multiple media platforms and changing social and political contexts (Sonnevend 2013b).

Let's take the example of the media representations of Steve Jobs's death in 2011:

1. *Foundation*: The coverage of this event built on the already-existing global iconic power of Steve Jobs and Apple Inc.
2. *Mythologization*: The event communicated a resonant message about a "genius" inventor who single-handedly changed our digital culture.
3. *Condensation*: For two weeks after Jobs's death, the Apple website summarized his image in a few clear, condensed sentences: "Apple has lost a visionary and creative genius, and the world has lost an amazing human being. Those of us who have been fortunate enough to know and work with Steve have lost a dear friend and an inspiring mentor. Steve leaves behind a company that only he could have built, and his spirit will forever be the foundation of Apple." This message of a "visionary and creative genius," who worked with relentless "passion," is also communicated by the often-reproduced iconic photographs of Steve Jobs in his trademark black turtleneck and blue jeans uniform. These images serve as lasting visual condensations of Apple's former CEO.
4. *Counternarration*: While Steve Jobs's life was celebrated worldwide, critical views on his leadership style and personality were also immediately shared, in efforts to counter the simplification and universalization of his legacy.
5. *Remediation*: Digital technologies enable faster international diffusion of global iconic events than ever before. Steve Jobs's life and death are now communicated in a myriad of media from Wikipedia sites to private memorials to Internet memes (Shifman 2013). His legacy is also rehearsed at Apple mega events that aspire to become preplanned global iconic events themselves.

Through these transnational storytelling practices, a global iconic event comes into being. Some global iconic events are more readily transportable than others, some have more counternarratives than others, but these five dimensions are generally present in their narration.

In sum, I have examined four features of "events in media:" (1) the power of the occurrence vis-à-vis its narrative as an "event," (2) the witnesses who tell the story of an "event," (3) the embodiments of the "event" in a variety of media, and (4) the travel of "events" across cultural and geographic boundaries. I have also tried to demonstrate that digital technologies make these features of "events in media" much more salient.

Occurrences and events exist in the digital age, as they have before and as they will after. Once the digital age becomes history, it too may be framed as an event, while modern humans experience a new, postdigital age, whose name we do not and cannot yet know. Understanding "events" thus helps us apprehend the "digital event" we are experiencing right now, in one moment of history. In other words, conceptualizing events is essential if we are to understand the digital condition itself.

See in this volume: archive, community, culture, digital, meme, memory

See in Williams: development, history, mediation, modern, myth, nature, tradition

References

Alexander, Jeffrey C. 2002. "On the Social Construction of Moral Universals: The 'Holocaust' from Mass Murder to Trauma Drama." *European Journal of Social Theory* 5(1): 5–86.

Balkin, Jack M. 2004. "Digital Speech and Democratic Culture: A Theory of Freedom of Expression for the Information Society." *New York University Law Review* 79(1): 1–55.

Blondheim, Menahem, and Tamar Liebes. 2002. "Live Television's Disaster Marathon of September 11 and Its Subversive Potential." *Prometheus* 20(3): 271–76.

Carey, James W. 1998. "Political Ritual on Television: Episodes in the History of Shame, Degradation and Excommunication." In *Media, Ritual and Identity*, edited by Tamar Liebes and James Curran, 42–70. London: Routledge.

Cottle, Simon. 2006. "Mediatized Rituals: Beyond Manufacturing Consent." *Media, Culture & Society* 28(3): 411–32.

Couldry, Nick, Andreas Hepp, and Friedrich Krotz, eds. 2010. *Media Events in a Global Age*. London: Routledge.

Dayan, Daniel, and Elihu Katz. 1992. *Media Events: The Live Broadcasting of History*. Cambridge, MA: Harvard University Press.

Frosh, Paul, and Amit Pinchevski. 2009. *Media Witnessing: Testimony in the Age of Mass Communication*. Basingstoke: Palgrave Macmillan.
Hallin, Daniel C. 1986. *"The Uncensored War": The Media and Vietnam*. New York: Oxford University Press.
Kampf, Zohar. 2009. "Public (Non-) Apologies: The Discourse of Minimizing Responsibility." *Journal of Pragmatics* 41(11): 2257–70.
Katz, Elihu, and Tamar Liebes. 2007. "'No More Peace!': How Disaster, Terror and War Upstaged Media Events." *International Journal of Communication* 1: 157–66.
Kellner, Douglas. 2003. *Media Spectacle*. London: Routledge.
Liebes, Tamar. 1998. "Television's Disaster Marathons: A Danger for Democratic Processes?" In *Media, Ritual and Identity*, edited by Tamar Liebes and James Curran, 71–84. London: Routledge.
Lull, James, and Stephen Hinerman, eds. 1997. *Media Scandals: Morality and Desire in the Popular Market Place*. Cambridge: Polity Press.
Neiger, Motti, Oren Meyers, and Eyal Zandberg. 2011. *On Media Memory: Collective Memory in a New Media Age*. New York: Palgrave Macmillan.
Peters, John D. 2001. "Witnessing." *Media, Culture & Society* 236: 707–23.
Pinchevski, Amit, and Tamar Liebes. 2010. "Severed Voices: Radio and the Mediation of Trauma in the Eichmann Trial." *Public Culture* 22(2): 265–91.
Scannell, Paddy. 1996. *Radio, Television and Modern Life*. Oxford: Blackwell.
———. 2014. *Television and the Meaning of 'Live': An Enquiry into the Human Situation*. Cambridge: Polity Press.
Schudson, Michael. 1992. *Watergate in American Memory: How We Remember, Forget, and Reconstruct the Past*. New York: Basic Books.
———. "The Anarchy of Events and the Anxiety of Story Telling." In *Why Democracies Need an Unlovable Press*, 50–63. Cambridge: Polity Press.
Sewell, William Jr. 1996. "Historical Events as Transformations of Structure: Inventing Revolution at the Bastille." *Theory and Society* 25(6): 841–81.
Shifman, Limor. 2013. *Memes in Digital Culture*. Cambridge, MA: MIT Press.
Sonnevend, Julia. 2013a. "Counterrevolutionary Icons." *Journalism Studies* 14(3): 336–54.
——— 2013b. "Global Iconic Events: How News Stories Travel through Time, Space and Media." PhD diss., Columbia University.
——— 2015. "'Symbol of Hope for a World without Walls': The Fall of the Berlin Wall as a Global Iconic Event." *Divinatio* 39–40: 223–33.
Sturken, Marita. 1997. *Tangled Memories: The Vietnam War, the AIDS Epidemic, and the Politics of Remembering*. Berkeley: University of California Press.
Thompson, John B. 2000. *Political Scandal: Power and Visibility in the Media Age*. Cambridge: Polity Press.
Wagner-Pacifici, Robin. 2010. "Theorizing the Restlessness of Events." *American Journal of Sociology* 115(5): 1351–86.
Zelizer, Barbie. 1992. *Covering the Body: The Kennedy Assassination, the Media, and the Shaping of Collective Memory*. Chicago: University of Chicago Press.

11

Flow

Sandra Braman

Old flow still flows.

—Charles Olson, *Notes on Poetics (towards Projective Verse II)*

Old words are good words. Large enough for all that might ensue.

Flow is an old word. In the Western tradition, there has been public betrayal of awareness that *Panta rhei*, "Everything flows," since at least the pre-Socratics. Other traditions document even longer histories. As happens to all old words, the meaning of *flow* changes over time and yet something constant, of its essence, always remains. This semantic drift within seeming stability, with its ebbs and eddies, is just the point. In the words attributed to Heraclitus, "for those who step into the same rivers, different and again different waters flow."

It is an old word, but flow is very much of the twenty-first century. A generation admonished to "go with the flow," as Fred Turner (2006) reminds us, built the Internet and is now concerned about whether all our information gets to flow the same way (network neutrality). Depending on scholarly predilection, one might be thinking about "places and flows" (Castells 2000) or the flows of "mediascapes" (Appadurai 1990). Certainly there are questions about who is making money from those flows of information, and how, and marketers—as will be further discussed below—are very much onto flow as the experience that will sell their products and services.

That does make *flow* an important keyword for the twenty-first century, so it is useful to focus on the enduring essence of the word as we explore its contemporary uses and manifestations. What remains the same, across time, cultures, disciplines, genres, and uses, is that flow—the exchange of information—is essential to the existence of systems. Flow therefore offers both liberatory and repressive opportunities, both distributive and capital-accumulating options. Culture, grace, and the most boorish instrumentalism.

Useful for those who do either administrative or critical research, using Lazarsfeld's (1948) still-influential categories, this systems approach also provides the theoretical foundation for a third type of research. Administrative research seeks to find ways of doing what is already being done better within a given system. The critical research that makes so much use of Williams's (1976) keywords looks for better ways of doing things within a given system or challenging an aspect of that system, sometimes to the point of revolution. Systems research tries to figure out just what the system actually is, knowing that any system involves far more, and often other, than what may be depicted on any map, chart, or table. The goal is to figure out just what it is that is being done, by whom, and how, in an environment assumed to be dynamic and to involve multiple systems interacting simultaneously but quite differently across the same and other scales.[1] Individual and group identities and agency—both as spent and as effective—emerge out of such knowledge. With this move, the degrees of freedom for policymakers and citizens, organizations and autonomous networks, vastly expand—thinking in systems terms thus inevitably brings power into the analysis.[2]

The political import of the word *flow* is evident in its history: It was in the late nineteenth century, according to the *Oxford English Dictionary*—just as the bureaucratic state was becoming formalized in much of Europe and North America—that the word *flow* came into use not only as the opposite of the "line" and the "path," both reverberations of the ancient word *logos* with which the Greeks opposed flow, but, also, the "fracture." The word was used in connection with electronics at least as early as the 1880s, when mathematicians started looking at unidirectional flows of electricity and magnetism through wires. Information theory (Shannon 1948) developed out of the effort to maximize the flow of voice telephone conversations through the AT&T telecommunications network.

What is at issue when we think about flow is not reducible to the opposition of stasis versus change, nor to that of being versus becoming. These nonanalogous oppositions can be engaged, even axially, but they do so while operating on their own, quite separate, trajectories; the latter cannot be reduced to the former unless analysis is unidimensional and applied to a single system operating as if in isolation. What is at issue is what and how much needs to flow in order for a given system to work, and what level and kind of stasis within or relative to that flow are necessary for identity and effective agency to emerge within that system as it navigates its relations with other systems at the same and other levels, of the same and other kinds.

A 2014 review of the contemporary literature on flow quickly finds a systems orientation in research across disciplines. It is there in studies of the neurological flows that bring about consciousness (e.g.,[3] León-Dominguez et al. 2013; Yanagawa et al. 2013). Systems are central to the concept of flow used in work on what it is that keeps users attracted to new technologies (Kim and Han 2014), on videogame addiction (Stavropoulos, Alexandraki, and Motti-Stefanidi 2013; Tokugawa 2011), and on users so engrossed in a website that they click "buy" (Hsu, Chang, and Chen 2012). Industries in which such studies have taken place include not only the expected social media (Kwak, Choi, and Lee 2014; Mauri et al. 2011) and games (Schmierbach et al. 2014), but also online banking (Zhou 2012). Implicit versions of systems-based approaches can be found in research on the flow of content through each stage of the entire production and consumption chain (Choi and Park 2014). There are studies of very specific types of intra-organizational flows (Detert et al. 2013; Erhardt, Martin-Rios, and Harkins 2014).

Flow isn't of interest only for commercial purposes. Choi (2014), for example, thinks about what it takes to create a public sphere via flows of political communication across online forums. The historic focus on flow in telecommunications regulation[4] is sustained in contemporary struggles over network neutrality, a debate over who should be allowed to "version" access to the network, and who can restrict access to certain types of information and content (Frieden 2008). Who should be allowed, that is, to create and enforce the informational and communicative class structures considered to have constitutional status in the United States (Braman 1989).

Game-theoretic modeling continues to be popular (Olaziola and Valenciano 2014), but we also now see work that combines flow theory with one or another version of contemporary systems theory in an effort to understand the nature and roles of communication in the face of disruptions (Ceja and Navarro 2012) or crisis (Baber et al. 2013). Diverse scenarios have stimulated analysis of flows in atypical, often turbulent or chaotic, environments—those in which it is to be expected that at least some of the dangers that must be addressed will be unfamiliar of kind. Thus we see analyses of flows during emergencies (Gao et al. 2014), in search-and-rescue operations (Baber et al. 2013), and in analysis of disorder in arenas as diverse as the natural ecosystems and children's problem solving (Castillo and Kloos 2013). Perhaps inspired—or, at least, foreseen—by former US Attorney General Ashcroft's insistence on the utility and validity of identity information generated by extrapolation in the search for imagined missing links in perceived flows, there is now research in this area as well (Feczak, Hossain, and Carlsson 2014).

Raymond Williams (1974) used the word *flow* to refer to streams of content on a channel across the course of a viewing period. This is now considered the "traditional" view of the (now traditional) medium as a stream of channel-specific content with a consistent identity built through scheduling structure and continuity and (ideally from the content producer's perspective) viewer experience (van den Bulck and Enli 2014). This understanding of flow remains a touchstone for communication scholars, but in the intervening forty years since Williams's book was published his referent has so changed that the type of flow he had in mind is barely cognizable. There are two problems. First, because of the robustness, over time and circumstance, of the finding that there can be quite significant differences in flow as observed and as perceived, Schumate and Contractor (2013) include that distinction as a fundamental dimension in their important analyses of flow. They argue that perceived flow, like perceived networks, should be treated analytically as subjects different in kind from those that are observed. That is, evidence of flow must be both conceptually and operationally distinguished from evidence of perceptions of flow. Second, most content is being, and has long been, designed with the expectation that it will subsequently be broken up into small pieces for redistribution through a variety of channels by diverse types of distributors (Baym and Shah 2011; Reeves 2013). The Williams-type flow is in the minds of one type of record-keeper regarding the content production and distribution enterprise only, and only one among many types of "users" or "viewers."

There have been other uses of the concept of flow among those who study information, communication, and culture. Political scientist Ithiel de Sola Pool, whose *Technologies of Freedom* (1983a) should still be required reading for anyone trying to understand the legal and policy side of the digital environment, developed an approach to studying international information flows (Pool 1983b) that remains in use today. Neuman and his colleagues recently looked at changes in the flow of information into the home over several decades (Neuman, Park, and Panek 2012), building on work that Neuman and Pool published in the mid-1980s. Other recent studies using this approach are diverse in terms of actual flows examined: Moon, Barnett, and Lim (2010) studied international music flows, for example, and Lotan et al. (2011) analyzed Twitter flows during political upheaval in North Africa. We continue to see analyses of the impact of international news flow content of the type that dominated during the New World Information Order (NWIO) debates of the late 1970s and 1980s (Avle 2011; Veltri 2012). Other types of research involving conceptualizations of flow that have been in use for decades include work on the ways in which social networks affect trade (Sangita 2013) and knowledge (Boussebaa, Sturdy, and Morgan 2014) flows.

Political scientist James Rosenau introduced the notion of "cascading interdependence" in 1984, speaking to a world in which those studying international relations were well aware of the salience of international information flows (Deutsch and Merritt 1979) and in which the regulatory ancestors of today's Internet policy debates were taking place under the rubric of "transborder data flow" (Branscomb 1983). Recent discussion about how to reimagine place relative to such global flows includes the reassertion and reinforcement of old ideas about space. Van Kempen and Wissink

(2014) argue for the continued importance of neighborhoods, for example, conceiving of them as "collections of hybrid nodes connecting to a multiplicity of flows binding actors and objects" (95). On the other hand, Hyman (2013) suggests decoupling resource flows from cities in their effort to rethink the urban environment. Prytherch (2010) revives the "communication as nervous system" analog by discussing the nodes of place as vertebral, and there are empirical studies of the spatial placement of cities relative to backbones such as that of the Internet and air traffic flows (Devriendt, Derudder, and Witlox, 2010).

Other concepts long in use are being revisited. Whelan et al. (2010), for example, expand on our understanding of how gatekeeping affects content and information flows when the Internet is involved. Strategic coupling, the process through which local assets are linked with global flows of network demands, can be reconsidered as capacity (Jacobs and Lagendijk 2014). News flow across the digital environment, from sources to aggregators like the "wire services" of old, to transfers across classes and genres of the news, can now be visualized through network analysis (Weber and Monge 2011). Such efforts may vary in their level of innocence; Harrison and Growe (2014) remind us that governments at various levels make good use of flow-driven concepts of regionalism for their own purposes as well.

Long-standing research methods for studying flows remain in use but are becoming further articulated. Sato (2014), for example, found that analysis of the carbon burden of international trade was much more useful if data were disaggregated down to the levels of individual products and the materials embodied within them. (The design of statistical categories is, of course, very much an information policy issue.) Cross-scale flows are of interest, whether those arise in the combination of international and locally intensified interactions involving an artisanal wine community (Mitchell et al. 2014), how particular North African films become available within immigrant communities in a city like Antwerp (Smets et al. 2013), or within what appear from the outside to be "merely" local environments (McIntyre et al. 2013).

Some of the lucidity and utility of earlier conceptualizations of the concept of flow for the purposes of research on information, communication, and culture have been lost in the rush toward abstractions at the level of "places and flows" (Castells 2000) or "mediascapes" (Appadurai 1990). It is not surprising that many researchers find it difficult to operationalize such ideas for research purposes. For authors such as Sutherland (2013), that is not a problem because the concept of flow is metaphysical and thus to be treated as a metaphor only, not appropriately the subject of empirical investigation. Working researchers, though, continue to push the development of flow conceptualization in their work on the digital environment.

A particularly well-articulated and useful contribution comes from a stream of work by Noshir Contractor and his many collaborators developing the theories of multidimensional and multiplex networks operationalized in their empirical research on networks and flows in a wide variety of types of systems. Network flow relations are defined as "patterns of message exchange or transmission among nodes" (Schumate and Contractor 2013, 456). In addition to the distinctions between observed and perceived flows discussed above, they distinguish between flows driven by individual motivations and those stimulated by joint goals, with further differentiations among types of joint goals possible. Aware of the same type of macro-level manifestations of globalization of interest to authors such as Castells and Appadurai, Contractor and his colleagues go much further and point out that not everything that happens globally, or across distance or through a network, should necessarily be conceived of as flow. Affinity relations are "socially constructed relationships among actors that may have either a positive or negative valence" (459). Representational relations are "affiliations among a set of actors that are communicated to a third party" (460). Semantic relations involve relationships among the data that flow through a network.

In a world in which we can, with straight faces, discuss "posthuman law" (Braman 2002), and in which object-oriented ontology (e.g., Harman 2011) is sweeping social theory with its focus on the "nonhuman," it is also worth noting that the conceptual framework being built by Contractor and his colleagues goes beyond the human as well to include objects and ideas among possible agents and factors. Seen through these lenses, much of what is referred to as "flow"—including by authors such as Castells and

Appadurai—should more accurately, and would more usefully, be thought of in other terms. Whether those are the terms as conceived of by Contractor and his colleagues or via other sets of categories, the important point is that flow is not all. It is one of several, perhaps many, in an array of types of relationships that characterize and have been made possible by today's global digital environment. All of these are worthy of research attention, but they require different methodological approaches and together require theoretical pluralism.

That flow is only one of the possible things that can happen digitally and/or globally does not diminish its importance as the subject of research and as something of which it is useful for us all to be aware—and self-aware, when we are examining the systems of which we ourselves are a part via our participation in their flows. Researchers typically think about how a particular research method will provide insights into media effects, but here effects can help us think about research design and method. Below a certain level of flow, effects—the system enabled—can no longer be sustained; thus a research agenda should look at ecologies of flows, and at decedent as well as emergent phenomena and processes within them. Because the value of flow varies by location of the link or node through which it passes, there are a number of strategic and tactical possibilities for any given context and any given player. From this perspective, political activists (whether in government or not), when working with flow (in government or out) would think about the point(s) at which, and in which flow(s), effective intervention of what type(s) can be levied. Citizens, scholars, and researchers would work on recognizing those systems that affect our lives, how we live within those systems, and just what it is that might be done—what kinds of agency, of power, might be available from various positions.

Vendors and website designers would, and do, seek to maximize user experience of flow to ensure that their goods and services are included in the personal ecologies of as many consumers as they can reach. There is a growing literature on the effort to develop and enhance the flow experience by product, services, and website designers, all of whom see its achievement as central to their marketing strategies. The concept of flow as used here falls in the line of work by Mihaly Csikszentmihalyi (1991), whose research on the experience of flow in, by now, thousands of people across cultures, ages, societies, jobs, and activities has consistently found that achieving the experience requires an intricate balance between complexity and skill, the determined and the indeterminate.

Many researchers now understand flow to be a phenomenon that appears far along on a spectrum of familiarity with, mastery of, and repetition of particular behaviors toward the pole of addiction, a perspective that again draws attention to the politics of flow and of research on it. This is an interesting moment in which the politics of flow could and should become an issue. It is plausible that parents and educators might draw the line between healthy flow and unhealthy addiction somewhat differently from how that is drawn by those selling products that rely upon such involvement for their success. For those of us on what might fashionably, today, be called the "user spectrum," these are also the moments in which we make our own political choices as we craft what Nardi and O'Day (1999) have so beautifully helped us understand as our personal information ecologies.

Here is where choice comes in. Ever rising, like a target. By engaging in particular flow and flows—particular by kind, by moment, and by context—by yielding to them, contributing, participating, we strengthen the system those flows enable and sustain. There are reasons why this old word, at many points in history, was considered synonymous with *morass*. Today the panic is as much about how much information is flowing (Hilbert and López 2011) as it is about what is being said.

The research, and the life, problem is to recognize the system that is being enabled by the flows in which we participate, or are being enticed to participate. Some may be for our pleasure; some may contribute to a rising political tide; some may be simply because we cannot stop. Whether as an individual, or a community, or a polity, however, we can and do make choices. The research question, then, is to very clearly identify in its articulated details the system(s) involved. These may or may not be as represented, claimed, referred to, expected, or imagined.

We are stepping into the same river again. They tell us it is overwhelming. We wonder. We step in.

See in this volume: algorithm, analog, archive, information, memory, mirror, personalization

See in Williams: alienation, capitalism, community, determine, ecology, organic, status, structural, unconscious

Notes

1 With bows to architectural theorist Sanford Kwinter (1992) for inspiring my own explorations of systems theory, and to communication scholar Klaus Krippendorff (1984/1993) for bringing complex systems theory to the field of communication early on, an introduction to the basics of complex adaptive systems theory, as they can be used to think about communication and about the democratic potential of the Internet, can be found in Braman (1994).

2 A qualitative increase in the flexibility and range of degrees of freedom also distinguishes informational metatechnologies—such as digital technologies and contemporary biotechnologies—from industrial technologies; see Braman (2004a).

3 Cites used here in this discussion of the literature are exemplars representing streams of work all worth including in a full-scale literature review of the kind for which there isn't sufficient space here.

4 For overviews of that history, see Braman (2004b) for interactions between law and policy for communication and information flow in the United States throughout its history, and Braman (1995) for a look at the run-up to today's Internet policy as it had unfolded for decades, issue by issue and technology by technology, up to the mid-1990s.

References

Appadurai, Arjun. 1990. "Disjuncture and Difference in the Global Cultural Economy." In *Global Culture: Nationalism, Globalization and Modernity*, edited by Mike Featherstone, 295–310. London: Sage Publications.

Avle, Seyram. 2011. "Global Flows, Media and Developing Democracies: The Ghanaian Case." *Journal of African Media Studies* 3(1): 7–23.

Baber, C., N. A. Stanton, J. Atkinson, R. McMaster, and R. J. Houghton. 2013. "Using Social Network Analysis and Agent-Based Modelling to Explore Information Flow Using Common Operational Pictures for Maritime Search and Rescue Operations." *Ergonomics* 56(6): 889–905.

Baym, Geoffrey, and Chirag Shah. 2011. "Circulating Struggle." *Information, Communication & Society* 14(7): 1017–38.

Boussebaa, Mehdi, Andrew Sturdy, and Glenn Morgan. 2014. "Learning from the World? Horizontal Knowledge Flows and Geopolitics in International Consulting Firms." *International Journal of Human Resource Management* 25(9): 1227–42.

Braman, Sandra. 2004a. "The Meta-technologies of Information." In *Biotechnology and Communication: The Meta-technologies of Information*, edited by Sandra Braman, 3–36. Mahwah, NJ: Lawrence Erlbaum Associates.

———. 2004b. "Where Has Media Policy Gone? Defining the Field in the Twenty-First Century." *Communication Law and Policy* 9(2): 153–82.

———. 2002. "Posthuman Law: Information Policy and the Machinic World." *First Monday*, http://www.firstmonday.org/ojs/index.php/fm/issue/view/152.

———. 1995. "Policy for the Net and the Internet." *Annual Review of Information Science and Technology* 30: 5–75.

———. 1994. "The Autopoietic State: Communication and Democratic Potential in the Net." *Journal of the American Society of Information Science* 45(6): 358–68.

———. 1989. "Information and Socioeconomic Class in US Constitutional Law." *Journal of Communication* 39(3): 163–79.

Branscomb, Anne Wells. 1983. "Global Governance of Global Networks: A Survey of Transborder Data Flow in Transition." *Vanderbilt Law Review* 36: 985–1043.

Castells, Manuel. 2000. *The Rise of the Network Society*. Vol. 1, *The Information Age: Economy, Society, and Culture*. 2nd ed. Cambridge, MA: Blackwell.

Castillo, Ramon D., and Heidi Kloos. 2013. "Can a Flow-Network Approach Shed Light on Children's Problem Solving?" *Ecological Psychology* 15(3): 281–92.

Ceja, Lucia, and Jose Navarro. 2012. "'Suddenly I get into the zone': Examining Discontinuities and Nonlinear Changes in Flow Experiences at Work." *Human Relations* 65(9): 1101–27.

Choi, Sujin. 2014. "Flow, Diversity, Form, and Influence of Political Talk in Social-Media-Based Public Forums." *Human Communication Research* 40: 209–37.

Choi, Sujin, and Han Woo Park. 2014. "Flow of Online Content from Production to Consumption in the Context of Globalization Theory." *Globalizations* 11(2): 171–87.

Csikszentmihalyi, Mihalyi. 1991. *Flow: The Psychology of Optimal Experience*. New York: HarperPerennial.

Detert, James R., Ethan R. Burris, David A. J. Harrison, and Sean R. Martin. 2013. "Voice Flows to and around Leaders: Understanding When Units Are Helped or Hurt by Employee Voice." *Administrative Science Quarterly* 58(4): 624–68.

Deutsch, Karl W., and Richard L. Merritt. 1979. "Transnational Communications and the International System." *Annals of the American Academy of Political and Social Science* 442(1): 84–97.

Devriendt, Lomme, Ban Derudder, and Frankl Witlox. 2010. "Conceptualizing Digital and Physical Connectivity: The Position of European Cities in Internet Backbone and Air Traffic Flows." *Telecommunications Policy* 34(8): 417–29.

Erhardt, Niclas, Carlos Martin-Rios, and Jason Harkins. 2014. "Knowledge Flow from the Top: The Importance of Teamwork Structure in Team Sports." *European Sport Management Quarterly* 14(4): 375–96.

Feczak, Szabolcs, Liaquat Hossain, and Sven Carlsson. 2014. "Complex Adaptive Information Flow and Search Transfer Analysis." *Knowledge Management Research & Practice* 12: 29–35.

Frieden, Rob. 2008. "A Primer on Network Neutrality." *Intereconomics* 43(1): 4–15.

Gao, Liang, Chaoming Song, Ziyou Gao, Albert-Lazlo Barbasi, James P. Bagrow, and Dashun Wang. 2014. "Quantifying Information Flow during Emergencies." *Scientific Reports* 4.

Harman, Graham. 2011. *The Quadruple Object*. Alresford, UK: Zero Books.

Harrison, John, and Anna Growe. 2014. "From Places to Flows? Planning for the New 'Regional World' in Germany." *European Urban and Regional Studies* 21(1): 21–41.

Hilbert, Martin, and Priscila López. 2011. "The World's Technological Capacity to Store, Communicate, and Compute Information." *Science* 332(6025): 60–65.

Hsu, Chia-Lin, Kuo-Chien Chang, Mu-Chen Chen. 2012. "Flow Experience and Internet Shopping Behavior: Investigating the Moderating Effect of Consumer Characteristics." *Systems Research and Behavioral Science* 29(3): 317–32.

Hyman, Katherine. 2013. "Urban Infrastructure and Natural Resource Flows: Evidence from Cape Town." *Science of the Total Environment* 461: 839–45.

Jacobs, Wouter, and Arnaud Lagendijk. 2014. "Strategic Coupling as *Capacity*: How Seaports Connect to Global Flows of Containerized Transport." *Global Networks* 14(1): 44–62.

Kim, Yoo Jung, and Jin Young Han. 2014. "Why Smartphone Advertising Attracts Customers: A Model of Web Advertising, Flow, and Personalization." *Computers in Human Behavior* 33: 256–69.

Krippendorff, Klaus. 1984/1993. "Information, Information Society and Some Marxian Propositions." In *Between Communication and Information: Information and Behavior*, edited by Brent Ruben and Jorge Schement, 4:487–521. New Brunswick, NJ: Transaction Books.

Kwak, Kyu Tae, Se Kyoung Choi, and Bong Gyou Lee. 2014. "SNS Flow, SNS Self-Disclosure and Post Hoc Interpersonal Relations Change: Focused on Korean Facebook User." *Computers in Human Behavior* 31: 294–304.

Kwinter, Sanford. 1992. "Landscapes of Change: Boccioni's *Stati d'animo* as a General Theory of Models." *Assemblage* 19: 50–65.

Lazarsfeld, Paul. 1941. "Remarks on Administrative and Critical Communications Research." *Studies in Philosophy and Social Science* 9(1–2): 2–16.

León-Domínguez, Umberto, Antonio Vela-Bueno, Manuel Froufé-Torres, and José León-Carrión. 2013. "A Chronometric Functional Sub-network in the Thalamo-cortical System Regulates the Flow of Neural Information Necessary for Conscious Cognitive Processes." *Neuropsychologia* 51(7): 1336–49.

Lotan, Gilad, Erhardt Graeff,, Mike Ananny, Devin Gaffney, Ian Pearce, and Danah Boyd. 2011. "The Revolutions Were Tweeted: Information Flows during the 2011 Tunisian and Egyptian Revolutions." *International Journal of Communication* 5: 1375–1405.

Mauri, Mauriozio, Pietro Cipresso, Anna Balgera, Marco Villamira, and Giuseppe Riva. 2011. "Why Is Facebook So Successful? Psychophysiological Measures Describe a Core Flow State While Using Facebook." *Cyberpsychology Behavior and Social Networking* 14(12): 723–31.

McIntyre, Julie, Rebecca Mitchell, Brendan Boyle, and Shaun Ryan. 2013. "We Used to Get and Give a Lot of Help: Networking, Cooperation and Knowledge Flow in the Hunter Valley Wine Cluster." *Australian Economic History Review* 53(3): 247–67.

Mitchell, Rebecca, Brendan Boyle, John Burgess, and Karen McNeil. 2014. "'You can't make a good wine without a few beers': Gatekeepers and Knowledge Flows in Industrial Districts." *Journal of Business Research* 67(10): 2198–2206.
Moon, Shin-Il, George A. Barnett, and Yon Soo Lim. 2010. "The Structure of International Music Flows Using Network Analysis." *New Media & Society* 12(3): 379–99.
Nardi, Bonnie A., and Vicki L. O'Day. 1999. *Information Ecologies: Using Technology with Heart*. Cambridge, MA: MIT Press.
Neuman, W. Russell, and Ithiel de Sola Pool. 1986. "The Flow of Communications into the Home." In *Media, Audience, and Social Structure*, edited by Sandra Ball-Rokeach and Muriel G. Cantor,71–86. Thousand Oaks, CA: Sage Publications.
Neuman, W. Russell, Yong Jin Park, and Elliot Panek. 2012. "Tracking the Flow of Information into the Home: An Empirical Assessment of the Digital Revolution in the United States, 1960–2005." *International Journal of Communication* 6: 1022–41.
Olaziola, Norma, and Federico Valenciano. 2014. "Asymmetric Flow Networks." *European Journal of Operational Research* 237(2): 566–79.
Pool, Ithiel de Sola. 1983a. *Technologies of Freedom*. Cambridge, MA: Belknap Press of Harvard University Press.
———. 1983b. "Tracking the Flow of Information." *Science* 221(4611): 609–13.
Prytherch, David L. 2010. "'Verbetrating' the Region as Networked Space of Flows: Learning from the Spatial Grammar of Catalanist Territoriality." *Environment and Planning A* 42(7): 1537–54.
Reeves, Joshua. 2013. "Temptation and Its Discontents: Digital Rhetoric, Flow, and the Possible." *Rhetoric Review* 32(3): 314–30.
Rosenau, James. 1984. "A Pre-theory Revisited: World Politics in an Era of Cascading Interdependence." *International Studies Quarterly* 28: 245–305.
Sangita, Seema. 2013. "The Effect of Disaporic Business Networks on International Trade Flows." *Review of International Economics* 21(2): 266–80.
Sato, Misato. 2014. "Product Level Embodied Carbon Flows in Bilateral Trade." *Ecological Economics* 104: 106–17.
Schmierbach, Mike, Mun-Young Chung, Mu Wu, and Keungyeong Kim. 2014. "No One Likes to Lose: The Effect of Game Difficulty on Competency, Flow, and Enjoyment." *Journal of Media Psychology* 26(3): 105–10.
Schumate, Michelle, and Noshir S. Contractor. 2013. "Emergence of Multidimensional Social Networks." In *The SAGE Handbook of Organizational Communication*, edited by Linda L. Putnam and Dennis K. Mumby, 449–74. Thousand Oaks, CA: Sage Publications.
Shannon, Claude E. 1948. "A Mathematical Theory of Communication." *Bell System Technical Journal* 27(3): 379–423.
Smets, Kevin, Iris Vandevelde, Philippe Meers, Roel Vande Winkel, and Sofie Van Bauwel. 2013. "Diasporic Film Cultures from a Multi-level Perspective: Moroccan and Indian Cinematic Flows in and towards Antwerp (Belgium)." *Critical Studies in Media Communication* 30(4): 257–74.
Stavropoulos, Vasilis, Kyriaki Alexandraki, and Frosso Motti-Stefanidi. 2013. "Flow and Telepresence Contributing to Internet Abuse: Differences according to Gender and Age." *Computers in Human Behavior* 29(5): 1941–48.

Sutherland, Thomas. 2013. "Liquid Networks and the Metaphysics of Flux: Ontologies of Flow in an Age of Speed and Mobility." *Theory, Culture & Society* 30(5): 3–23.

Taylor, Peter J., Michael Hoyler, and Raf Verbruggen. 2010. "External Urban Relational Process: Introducing Central Flow Theory to Complement Central Place Theory." *Urban Studies* 47(13): 2803–3818.

Tokunaga, Robert Shota. 2011. "Engagement with Novel Virtual Environments: The Role of Perceived Novelty and Flow in the Development of the Deficient Self-Regulation of Internet Use and Media Habits." *Human Communication Research* 39: 365–93.

Turner, Fred. 2006. *From Counterculture to Cyberculture*. Chicago: University of Chicago Press.

Van den Bulck, Hilde, and Gunn Sara Enli. 2014. "Flow under Pressure: Television Scheduling and Continuity Techniques as Victims of Media Convergence?" *Television & New Media* 15(5): 449–52.

van Kempen, Ronald, and Bart Wissink. 2014. "Between Places and Flows: Towards a New Agenda for Neighborhood Research in an Age of Mobility." *Geografiska Annaler Series B: Human Geography* 96(2): 95–108.

Veltri, Giuseppe Alessandro. 2012. "Information Flows and Centrality among Elite European Newspapers." *European Journal of Communication* 27(4): 354–75.

Weber, Matthew S., and Peter Monge. 2011. "The Flow of Digital News in a Network of Sources, Authorities, and Hubs." *Journal of Communication* 61(6): 1062–81.

Whelan, Eoin, Robin Teigland, Brian Donnellan, and Willie Golden. 2010. "How Internet Technologies Impact Information Flows in R&D: Reconsidering the Technological Gatekeepers." *R&D Management* 40(4): 400–413.

Williams, Raymond. 1976. *Keywords: A Vocabulary of Culture and Society*. London: Croom Helm.

———. 1974. *Television: Technology and Cultural Form*. New York: Routledge.

Yanagawa, Toru, Zenas C. Chao, Naomi Hasegawa, and Naotaka Fujii. 2013. "Large-Scale Information Flow in Conscious and Unconscious States: An ECoG Study in Monkeys." *PLOS One* 8(11): e80845. doi:10.1371/journal.pone.0080845.

Zhou, Tao. 2012. "Examining Mobile Banking: User Adoption from the Perspectives of Trust and Flow Experience." *Information Technology & Management* 13(1): 27–37.

12

Forum

Hope Forsyth

Two thousand years of history, legislation, applications, and context converge in the term *forum*. *Forum*'s etymological basis and history set a foundation for considering it as a spatially grounded, physically embodied place of action, gathering, and societal interaction. Digital media have further complicated the term, with online bulletin boards and comment sections referred to as forums, while physical and commercial spaces have responded complementarily to these advances. Ultimately forums, whether spatial or digital, must be understood through physical and material terms.

From Trees to Gladiatorial Marches: Historical Context and Comparison

Forum is linguistically derived from the Latin *fores*, meaning "what is out of doors." Similar to *forest*, forum originally designated an outside space. Its meaning evolved during Roman rule into a type of liminal space within society but outside of one's home. In this context, a forum set apart a section of the community, neither home nor alien, as space where public and private concerns could be tended and civic gatherings could be held. Architectural historian David Watkin describes the Roman Forum, the best-known of the ancient forums, located between the Capitoline and Velian hills in Rome, as the "juridical, administrative, and commercial centre of Republican Rome [which] later became the key symbol of Roman imperial power" (2). The Forum served only those with specific credentials (namely, Roman citizens); accessibility was a key consideration in forums from early on. Additionally, the Forum intermingled business and military affairs, criminal trials and victory marches, religious shrines and the Senate.

Roman citizens' active societal lives were generally contained in the expanses of the uppercase-*F* Forum. In addition to the Forum's civic infrastructure—places for courts, sacrifices, military demonstrations, and so forth—a key feature was its spatial, physically embodied, human-supporting infrastructure, such as food stalls (Watkin 20) and a sewage system (*Ancient History Encyclopedia*), used to drain the swampy land the Forum occupied. In this way, the human-supporting infrastructure preceded and accompanied the civic infrastructure; without the former, physical constraints would have rendered the latter unsustainable.

One thing the Forum did *not* include was spectator sports. Those, including the most famous gladiatorial games, took place at arenas such as the Colosseum. Arenas, defined by the *Oxford English Dictionary* (*OED*), can be "any sphere of public or energetic action" but more primarily are "a scene or sphere of conflict; a battle-field." Citizens gathered in arenas for the largely passive activities of observation and entertainment. By contrast, the Forum provided for citizens a physical space for active gathering, participation, and progress.

Vestiges of the physical Roman Forum are still present today, as seen in the *OED*'s several definitions, including "as the place of public discussion" and "a court, tribunal." The first definition has found new life with the rise of online message boards, but the second is also curiously relevant. Legal precedent in the United States also considers forums in the realm of free speech and commerce, with the Supreme Court utilizing the word *forum* when delineating three distinct areas with First Amendment protections.[1] These spaces are distinguished by their accessibility and thus continue in the historical tradition of forums as liminal spaces. Forums serve communities and societies, but their liminal characteristics prevent them from being called public.

Considered generally, forums, whether uppercase-*F* Roman or lowercase-*f* American, exhibit three consistent attributes. First, they are spaces of societal gathering, where norms are reconfirmed and personal interests, including commerce and religion, are pursued. Second, they are liminal spaces of action, where both personal and public business is conducted. Third, they are physically embodied; the societal infrastructure[2] they provide is impossible without physical, and more specifically human-supporting, infrastructure designed to provide the basic physical necessities—such as shelter, food, and plumbing—of those acting within them.

Don't Read the Comments: Internet "Forums"

The internet resembles the Roman Forum in that it provides close and quick access to gathering and action, such as commerce, legislation and law enforcement, and religion. Many such activities—bill paying, tax filing, devotional reading, and news browsing—belong to private business, although public actions, such as searching open records or rallying community support, are not uncommon.

At the same time, online sites can have a tendency to resemble an arena more than a forum, especially given the plethora of opportunities for passive spectating. Flame wars and online comment sections can be as seductive as a car wreck in commandeering attention. Those who follow such clashes resemble the spectators in the Colosseum watching gladiatorial battles far more than citizens conducting business in the Forum. It's a peculiar perk (and drawback) that the internet can, with rapid-fire typing and clicking, switch back and forth between mimicking forums and mimicking arenas. Further, while a person can physically inhabit only one space at a time, an internet user can mingle among multiple online spaces at the same time.

Online message boards, commonly called forums, fulfill the first and second attributes of forums—public gathering and meaningful action—and, at first glance, appear to fulfill the third as well. After all, the internet cannot exist without its profoundly material infrastructure of physical wires, plugs, pixels, fiber-optic cables, displays, electricity grids, and sundry other material supports. Further, the internet does supply various types of human support, especially related to emotions and reputations.

Nevertheless, the mediated interactions and feedback systems of online forums are insufficient to qualify the digital gathering spaces in question as true forums under the present definition. Though they provide a wealth of resources and meaningful interactions, internet forums do not provide sufficient human-supporting infrastructure. Put flippantly, some types of hardware may be called chips, but they're no substitute for potatoes. John Perry Barlow triumphantly proclaimed in his "Declaration of the Independence of Cyberspace" that the internet "consists of transactions, relationships, and thought itself . . . [it] is a world that is both everywhere and nowhere, but it is not where bodies live." True enough, except that bodies have to live somewhere offline. Refocusing concentration not on internet "forums," but on the internet more broadly leads toward a potential resolution.

Sum of the Parts: Human Infrastructure and Action

The Roman Forum's evolution caused distinctly stratified layers of development. When fires raged or buildings became decrepit, new infrastructure sprang up on top of its predecessors. In a similar way, the evolution of the internet has prompted new layers of forums to develop, and just as the Forum's layers built upon each other, the internet provides yet another layer to the ongoing media archaeology propping up the modern-day forum.

Internet users have come to seek the missing human-supporting infrastructure of their media in order to have the most fruitful and productive interactions with it. Writing in "Where Code Meets Place," design scholar Laura Forlano emphasizes the interconnectedness of place and organization when people interact with the internet: "Physical spaces are quickly being mapped, located and layered with an invisible digital skin signaling a merger between the digital and the real, offline, analog worlds" (3). As computers have grown smaller, users' abilities to mesh interactions with the internet and interactions with the physical world into symbiotic experiences have increased. Though some might argue that this meshing succeeds a bit too well for comfort,[3] these interactions have the potential to build and rebuild forums in the coupling of gathering and action, whether online or off, within space and time.

Others have examined more closely one historical example of such a site of digital and analog mergers: namely, the coffee shop. Brian Cowan, a historian of early modern British history, establishes a historical context and precedent for considering the coffee shop as a significant physical space for social activism and public participation, in his 2005 book *The Social Life of Coffee*. He writes:

"The coffee house has been understood to be a novel and unique social space in which distinctions of rank were temporarily ignored and uninhibited debate on matters of political and philosophical interest flourished" (2). In 2003, the *Economist* published an article entitled "The Internet in a Cup: Coffee Fuelled the Information Exchanges of the 17th and 18th Centuries," which traces the development of coffee shops as "information exchanges for writers, politicians, businessmen and scientists" and remarks that "coffee-houses provided a forum for education, debate and self-improvement." Coffee shops became one of the first early modern places where gathering, interaction, and activity could occur outside of government-owned property or private homes.

Current coffee shops, from behemoth Starbucks to twee hipster hot spots, may not always host the lofty philosophical debates that Cowan and his predecessor Jürgen Habermas describe in their works. Rather, Wi-Fi-hospitable coffee shops are a fitting example of physical infrastructure and internet activity coupling to provide all three attributes of a forum. Their variety of gatherings and actions, from job interviews to Wi-Fi workers, and their necessary physical support structures—food, shelter, climate control, plumbing, and abundant caffeine—provide all three forum attributes. Coffee shops (and by extension, other Wi-Fi hot spot hosts), considered purely physically, provide the human-supporting infrastructure the internet needs to be a sustainable forum.[4]

Coffee shops function similarly to the Roman Forum in the way they establish liminal spaces, outside both the private domestic sphere of one's home and the public sphere of one's business, but still within society. Former Starbucks CEO Jim Donald explicitly reveals this motivation: "It's all part of their strategy to make Starbucks a third primary 'place' in the day of Americans. 'We say the first place is home, second place is office, and then Starbucks is a third place. . . . They use our stores for gathering spots, and we think that that's what makes that whole experience what it is today'" ("Starbucks' Psychology"). Perhaps the description was just meant as good marketing, but whether Donald intended to or not, through this idea of a "third place" between private (home) and public (office), he situates his company in the discourse of liminal spaces.[5]

Historical context establishes the forum as a site of physical embodiment, gathering, and action, whereas the arena appears as a site for spectating, entertainment, and gawking. Though *forum* has a number of senses, including judicial and historical, it is defined most clearly and generally by three attributes: gathering, action, and human-supporting infrastructure. Even with material aspects, the internet does not function as a sustainable forum considered on its own because of its lack of human-supporting infrastructure. Coffee shops—which first became sites of gathering and action in early modern Britain—now couple physical resources with internet accessibility, meeting all three attributes necessary to activate a new sense of a sustainable forum simultaneously online and off. In an inversion from its etymological roots, *forum* has evolved from being a descriptor of outside status to a mediator of inside and outside—a mediator that must balance accessibility and restriction, publicity and privacy, society and government. Through the coupling of infrastructure and internet, Silicon Valley innovations and one's favorite coffee are integrated into the ancient Roman Forum's strata.

See in this volume: analog, community, digital, internet, participation, sharing

See in Williams: city, community, culture, democracy, popular, private, regional, structural

Notes

1 During the 1980s and in response to multiple lawsuits regarding free speech in public schools, the Supreme Court established three types of forums where First Amendment rights to free speech can be exercised: public forums (i.e., "public parks, sidewalks and areas that have been traditionally open to political speech and debate"), which are open to all at all times; designated forums (i.e., "municipal theatres and meeting rooms at state universities"), which are used at specific times; and nonpublic forums (i.e., "airport terminals and a public school's internal mail system"), which require vetting ahead of participation (Cornell Law School). This delineation is mutually exclusive.

2 See Bowker and Star's *Sorting Things Out*, which identifies some features of infrastructure, including embeddedness in social arrangements and links with conventions of practice (Bowker and Star, 35).

3 Case in point: the person on holiday at the beach, smartphone in hand.

4 The *Economist* remarks characteristically that "history provides a cautionary tale for those hotspot operators that charge for access . . . information, both in the 17th century and today, wants to be free—and coffee-drinking customers, it seems, expect it to be."

5 The physical infrastructure of coffee shops goes beyond fulfilling the third criterion for a forum and extends influentially to another attribute: the action that takes place. The hum of a bustling coffee shop—espresso machines whistling, chairs scratching, greetings being exchanged, muted conversations, carefully cultivated indie music playing in the background—has been shown to enrich the creativity of the people within the space. A study done by several business administration professors on ambient noise concluded that "a moderate (vs. low) level of ambient noise is likely to induce . . . processing difficulty, which activates abstract cognition and consequently enhances creative performance. A high level of noise, however, reduces the extent of information processing, thus impairing creativity" (Mehta, Zhu, and Cheema, 785). In other words, working in a physically bustling forum improves the quality of that work—at least when it comes to creativity—by forcing the brain to create work-arounds and responses to small difficulties. This resembles an inoculation: prompt a response on a small scale and you get a productive result; mess with it a lot and things go haywire.

Forlano also recognizes this effect and remarks that "mobile professionals working in cafes are often surrounded by the loud screeching of the espresso machine. . . . While this could be seen as an inconvenience, many mobile workers report that sound is an important stimulant for their work" (155). This effect can be approximated somewhat with looped audio tracks of coffee shop sound, though the effect isless pronounced and the coffee less abundant; here again, physical infrastructure cannot be escaped. In this way, the human-supporting infrastructure and the spatial qualities of the coffee shop both complete the requirements of a forum and go beyond mere sustainment of the environment, enriching the other forum attributes and positively influencing the actors within the system.

References

Ancient History Encyclopedia. "The Roman Forum." http://www.ancient.eu/search/?q=the+roman+forum&sa.x=0&sa.y=0.

Barlow, John Perry. "A Declaration of the Independence of Cyberspace." https://projects.eff.org/~barlow/Declaration-Final.html.

Bowker, Geoffrey C., and Susan Leigh Star. *Sorting Things Out: Classification and Its Consequences*. Cambridge, MA: MIT Press, 1999.

Cornell Law School Legal Information Institute. "Forums." https://www.law.cornell.edu/wex/forums.

Cowan, Brian. *The Social Life of Coffee*. New Haven, CT: Yale University Press, 2011.

Forlano, Laura. "Where Code Meets Place." PhD diss., Columbia University, 2008. http://pqdtopen.proquest.com/doc/304625179.html?FMT=ABS.

"The Internet in a Cup." *Economist*, December, 2003. http://www.economist.com/node/2281736.

Mehta, Ravi, Rui (Juliet) Zhu, and Amar Cheema. "Is Noise Always Bad? Exploring the Effects of Ambient Noise on Creative Cognition." *Journal of Consumer Research* 39(4) (December 2012): 784–99. http://www.jstor.org/stable/10.1086/665048.

"Starbucks' Psychology." *ABC News*. http://abcnews.go.com/WN/Consumer/story?id=3162590.

Watkin, David. *The Roman Forum*. Cambridge, MA: Harvard University Press, 2009.

13

Gaming

Saugata Bhaduri

Gaming is generally understood, especially in the current context, as the act of playing video games or games with a digital interface. Accordingly, it is often erroneously presumed that while the use of *game* as a noun can be traced far back in the history of the English language, *game* as a verb and *gaming* as its present participle are of fairly recent origin. However, the *Merriam-Webster Dictionary*, among others, traces the use of *gaming* to the end of the fifteenth or the beginning of the sixteenth century, connecting this participial form etymologically to *gambling*. The *Online Etymology Dictionary* traces the word *game* as follows:

> Old English gamen "game, joy, fun, amusement", common Germanic (cognates: Old Frisian game "joy, glee", Old Norse gaman, Old Saxon, Old High German gaman "sport, merriment", Danish gamen, Swedish gamman "merriment"), regarded as identical with Gothic gaman "participation, communion", from Proto-Germanic *ga-collective prefix +*mann "person", giving a sense of "people together".

There are thus two important components that make up the sense of a word like *game*—an original sense of communion and a derivative sense of enjoyment. The present participial form *gaming* also includes a signal sense of taking risks or wagering, etymologically connected as it is to *gambling*. So gaming should be understood not as simply playing games but as entailing the three features of collectivization, enjoyment, and excess—the former two being already entailed in the word *game* itself, and the last coming into play in the present participial form of the digital keyword *gaming*.

While the word *gaming* has been available since the end of the fifteenth century, it appears to have been rarely used. Why is this so? How can one analyze the reticence of the English language to deploy the verbal—and more specifically, the present participial, and thus grammatically continuing and open-ended, never-foreclosed and never-ending—form of a word whose nominal form *game* has long been so frequent? In this essay I suggest both the liberatory and the transgressive quality unleashed in the recent rise of *gaming* as a gerund. While *game* itself was rooted in communion and enjoyment, the excess of the suffix *-ing* introduces the risqué and risky. Perhaps gam*ing* is thus essentially subversive, connected to the dangerous wastefulness of gambling. A closer look at the disruptive and myriad qualities of the modern gaming universe, including frequent role-playing and identity negotiation, will shed light on the dual nature, the promise and the peril, of modern gaming.

But before one can move to the present participial form *gaming* and its implications, let us take a look at what *game* itself denotes. There are two, presumably contradictory, elements that make up a game. On one hand, a game has to have a structure, fairly set rules, and definable goals and objectives; on the other, a game is supposed to lead to enjoyment. It is in this duality then that the primary feature of a game lies: it cannot be fully de-structured, or de-structive, given its structured set of rules and goals; and yet the fact that a game is also meant to produce joy or enjoyment challenges to disrupt that set structure. Several scholars have noted this instability: For instance, in his *Philosophical Investigations* (1953, 66–70), Wittgenstein finds *games* undefinable, while Lyotard and Thébaud too highlight the indeterminacy of gaming in their *Just Gaming* (1985). Games contain this tension, and the participial form *gaming* perpetually extends and defers the completion of a game; in gaming, the set structure of a game has not yet arrived.

To understand what *gaming* is all about, it may be worthwhile to compare and contrast it with a presumably similar word like *play*. Johan Huizinga's 1938 classic *Homo Ludens* posits that to play is what it means to be human—and the basis for human culture, including language, law, war, knowledge, poetry, philosophy, and art. Huizinga says that "culture arises in the form of play, that it is played from the very beginning" (46). In fact, Huizinga continues,

"Play is older than culture . . . and animals have not waited for man to teach them their playing" (1). The institution of play is both the source and the original form of human civilization: "We have to conclude, therefore, that civilization is, in its earliest phases, played. It does not come *from* play like a baby detaching itself from the womb: it arises *in* and *as* play, and never leaves it" (173). Like the unsettled structure of gaming, play, Huizinga suggests, at once demands and creates order, and is also the means to freedom itself (8–10). While Huizinga uses the word *play* and not *game* throughout, we may see games, gaming, and play as shared human forms for working out order and excess.

On the other hand, the differences hidden in the supposed cognates *game* and *play*, once examined, help shed light on what is distinctive about *gaming*. To define *play*, one can turn again to Huizinga and his classic definition:

> Summing up the formal characteristic of play, we might call it a free activity standing quite consciously outside "ordinary" life as being "not serious" but at the same time absorbing the player intensely and utterly. It is an activity connected with no material interest, and no profit can be gained by it. It proceeds within its own proper boundaries of time and space according to fixed rules and in an orderly manner. It promotes the formation of social groupings that tend to surround themselves with secrecy and to stress the difference from the common world by disguise or other means. (13)

Now contrast this fairly constrained definition of *play* with the tactical sense of the gaming term *gameplay*. *Gameplay* refers to the interactive and strategic experience of playing a game, or, as Craig Lindley puts it, "gameplay [is] a pattern of interaction with the game system. . . . In general, it is a particular way of thinking about the game state from the perspective of a player" (2004, 186). "The experience of gameplay" he continues elsewhere "is one of interacting with a game design in the performance of cognitive tasks, with a variety of emotions arising from or associated with different elements of motivation, task performance and completion" (Lindley, Nacke, and Sennersten 2008, 9). While play for Huizinga was absorbing, orderly, and constrained, gameplay for the gaming world involves openly strategic, individual, and contingent action. Gameplay introduces, as Gonzalo Frasca (2003) points out, "manipulation rules," or what an individual player can play beyond the set "goal rules" and "meta-rules" of the game (231–32). That manipulation and strategy are unique to gaming is further clarified by the word *gamesmanship*, or the art of strategically manipulating rules to win a game. This differs not only from the more orderly notion of play in Huizinga; it also differs from the sense of sportsmanship, as defined by Stephen Potter (1947)—playing by the rules and accepting defeat with grace—thus marking *gaming* as different from another presumably cognate word, *sports*, too.

Here we can see what may be the essential difference between games, on the one hand, and sports and play on the other. Games, and especially the present-continuous-tense sense of gaming (connected etymologically as it is to gambling), feature the possibility of subversion, strategy, and manipulation, while sports and play invoke mostly rule-bound action. The phrase *gaming the system* affirms this observation—as does the related and recent process of gamification, whereby gaming elements, like scoring points and competition, are extended to nongaming contexts like education, business, and online marketing. In this and other forms of gaming, including the interdisciplinary study of gaming theory, game studies, or ludology, the point is to find enjoyment for the player by reaching beyond the ordinary bounds of play, while also looking at the social implications of the same. It should not surprise that the academic and public reception of games, since at least the Frankfurt School, has at times raged against its excesses with concerns about the negative effects of graphic violence in games, the promotion of digital and class divides across the world, and many other social issues. Indeed, the gaming world, a far larger topic than I can take up in this conceptually narrow keyword study, courses with much of the vicious hatred and abuse that typify modern problems of sexism, racism, and classism. This essay neither denies nor engages with these gaming problems; rather it simply frames them as a deeply unfortunate consequence of the fundamentally transgressive character of gaming.

To move ahead with the relation between *game* and *gaming*, which this article believes to be the crux of understanding the latter as a keyword, one can see how game theory—initiated by John von Neumann in his 1928 article "On the Theory of Games of Strategy" and popularized in his coauthored 1944 book *Theory of Games and Economic Behaviour*—offers an interesting contrast case to gaming theory. Game theory concerns "the study of mathematical models of conflict and cooperation between intelligent rational decision-makers" (Myerson 1991, 1), and its key criteria of rationality, predictability, and determinability appear to diverge sharply from the subversive or excessive quality of gaming. Eric Rasmusen expands on the game in game theory:

> The essential elements of a game are players, action, payoffs, and information—PAPI, for short. These are collectively known as the rules of the game, and the modeller's objective is to describe a situation in terms of the rules of a game so as to explain what will happen in that situation. Trying to maximize their payoffs, the players will devise plans known as strategies that pick actions depending on the information that has arrived each moment. The combination of strategies chosen by each player is known as the equilibrium. Given an equilibrium, the modeller can see what actions come out of the conjunctions of all the players' plans, and this tells him the outcome of the game. (2001, 31–32)

Here we see again the earlier tension of all games. A PAPI-based model or approach to games theorizes games with the goal of successfully predicting, and thus eliminating uncertainty, the outcomes of games given certain strategies and conditions. Equilibrium is a far cry indeed from the constantly unsettled sense of *gaming*. Yet game theory, an artifact invented in the turbulent interwar period, is also an exacting response to and embrace of these same profound uncertainties. The creative uncertainty with which every act of gaming throws a player off balance is the same set of uncertainties game theory seeks to operationalize and bring back into balance. So, whether one considers it the antithesis of gaming theory or its natural consequence, or both, game theory—the risk-averse darling child of modern economics, political science, evolutionary biology, and some forms of pragmatist philosophy—may be distinguished from the openly risky senses of *gaming*.

Now let us consider digital games and gaming in particular. A cursory glance at the history of video games and digital gaming suggests a similar arc from relatively closed, nonmanipulable, single-player video games to the relatively recent development of open, user-adaptable, multiple-player, interactive role-playing, simulation gaming (with mods and hacks coursing throughout the gamer community). In a phrase: from Space Invaders to Second Life. Of course online gaming took root as early as 1974 with the rise of Mazewar, MUDs (multiuser dungeons), and MMORPGs (massively multiple-player online role-playing games), having built on an ancient tradition of offline gaming. And of course some of the most popular digital games—particularly the app-based social games popular on Zynga—have connected over a billion people online in what many consider mindless, gold-farming games no more subversive than Pong. The difference in the histories of popular video gaming, such as that by Steven Kent (2001), and the histories of online gaming, even as early as T.L. Taylor's (2006), evidences an uneven rise in collective open indeterminacy among digital gaming communities separate from closed app-based social games.

The collective creation of gaming subcultures is just as essential as enjoyment and excess to understanding modern gaming. A lone gamer enters a gaming community, according to McKenzie Wark (2007), through creative and subversive appropriation of societal norms, through subverting or rearranging official game narratives through hacking, modding, cheating, sandboxing, participating in or designing open world games, and through extending his or her own gameworld to other worlds through cosplay, fanfiction, machinima, and the like. In each of these three acts, gaming becomes a collective act of tactical reappropriation and creativity, excess and subversion. David Getsy (2011) too has pointed out how games—originally created for diversion—are first and foremost sites of subversion. There are many critical reflections that must be examined in contemporary gaming culture and subcultures—no less than the reifying and fantasizing of the virulently toxic politics of gender, race, and class discrimination that many scholars have analyzed.[1] These pressing issues deserve attention, although they fall outside the scope of this immediate keyword study. Rather, it may be fruitful to argue that perhaps the social problems that currently beset the gaming world have emerged, as indeed have its transgressive and enabling aspects, out of the consequence of the thesis of this essay: that twined excessive character of gam*ing*, and its ongoing openness to both virtue and vice, that ever renders our nominal world as a game into something subversive and unstable.

See in this volume: community, geek, hacker, meme, personalization, sharing

See in Williams: community, creative, folk, mechanical, myth, popular, violence

Note

1 For critical writings on gaming, including those exposing the problematic politics of gaming in the domains of race, class, gender, sexuality, etc., one can look at the following:

Select articles from readers and anthologies: e.g., David G. Embrick, J. Talmadge Wright, and Andras Lukacs, eds., *Social Exclusion, Power, and Video Game Play: New Research in Digital Media and Technology*(Lanham, MD: Lexington Books, 2012); Nina Huntemann, ed., *Joystick Soldiers: The Politics of Play in Military Video Games* (New York: Routledge, 2010); Frans Mäyrä, *An Introduction to Game Studies: Games in Culture* (London: Sage Publications, 2008); Ken S. McAllister, *Gamework: Language, Power, and Computer Game Culture* (Tuscaloosa: University of Alabama Press, 2004); Jason C. Thompson and Marc A. Ouellette, eds., *The Game Culture Reader* (Newcastle upon Tyne: Cambridge Scholars Press, 2013); Mark J. P. Wolf and Bernard Perron, eds., *The Video Game Theory Reader* (London: Routledge, 2003).

Individual articles: e.g., C. A. Anderson and B. J. Bushman, "Effects of Violent Video Games on Aggressive Behavior, Aggressive Cognition, Aggressive Affect, Physiological Arousal, and Prosocial Behavior: A Meta-analytic Review of the Scientific Literature," *Psychological Science* 12(5) (2001): 353–59; D. A. Gentile, P. J. Lynch, J. R. Linder, and D. A. Walsh. "The Effects of Violent Video Game Habits on Adolescent Hostility, Aggressive Behaviors, and School Performance," *Journal of Adolescence* 27 (2004): 5–22; Nina Huntemann, "Introduction: Feminist Discourses in Games/Game Studies," *Ada: A Journal of Gender, New Media, and Technology* 2 (2013); Ewan Kirkland, "Masculinity in Video Games: The Gendered Gameplay of Silent Hill," *Camera Obscura* 24(2 71) (2009): 161–83; David J. Leonard, "Not a

Hater, Just Keepin' It Real: The Importance of Race- and Gender-Based Game Studies," *Games and Culture* 1(1) (2006): 83–88; T. L.Taylor, "Multiple Pleasures: Women and Online Gaming," *Convergence: The International Journal of Research into New Media Technologies* 9(1) (2003): 21–46.

References

Frasca, Gonzalo. 2003. "Simulation versus Narrative: Introduction to Ludology." In *The Videogame Theory Reader*, edited by Mark J.P. Wolf and Bernard Perron, 221–35. New York: Routledge.

Getsy, David J. ed., 2011. *From Diversion to Subversion: Games, Play and Twentieth-Century Art*. University Park: Pennsylvania State University Press.

Huizinga, Johan. 1938/1955. *Homo Ludens: A Study of the Play-Element in Culture*. Translated by C. Van Schendel. Boston: Beacon Press.

Kent, Steven L. 2001. *The Ultimate History of Video Games: From Pong to Pokémon and Beyond—The Story behind the Craze That Touched Our Lives and Changed the World*. New York: Three Rivers Press.

Lindley, Craig. 2004. "Narrative, Game Play, and Alternative Time Structures for Virtual Environments." In *Technologies for Interactive Digital Storytelling and Entertainment: Second International Conference TIDSE 2004, Darmstadt, Germany, June 2004, Proceedings*, edited by Stefan Göbel et al., 183–94. Berlin & Heidelberg: Springer Verlag.

Lindley, Craig, Lennart Nacke, and Charlotte Sennersten. 2008. "Dissecting Play—Investigating the Cognitive and Emotional Motivations and Affects of Computer Gameplay." In *CGAMES 08: Proceedings of 13th International Conference on Computer Games*, 9–17. Wolverhampton, UK: University of Wolverhampton. http://www.academia.edu/365971/Dissecting_Play-Investigating _the_ Cognitive_and_Emotional_Motivations_and_Affects_of_Computer _Gameplay.

Lyotard, Jean-François, and Jean-Loup Thébaud. 1985. *Just Gaming*. Translated by Wlad Godzich. Minneapolis: Minnesota University Press.

Merriam-Webster Dictionary. http://www.merriam-webster.com/dictionary/gaming.

Myerson, Roger B. 1991. *Game Theory: Analysis of Conflict*. Cambridge, MA: Harvard University Press.

Neumann, John von. 1928/1959. "On the Theory of Games of Strategy." Translated by Sonya Bargmann. In *Contributions to the Theory of Games*, edited by A.W. Tucker and R. D. Luce, 4: 13–42. Princeton, NJ: Princeton University Press.

Neumann, John von, and Oskar Morgenstern. 1944/1953. *Theory of Games and Economic Behaviour*. Princeton, NJ: Princeton University Press.

Online Etymology Dictionary. http://www.etymonline.com/index.php?term=game &allowed _in_frame=0.

Potter, Stephen. 1947. *The Theory and Practice of Gamesmanship: The Art of Winning Games without Actually Cheating*. London: Rupert Hart-Davis.

Rasmusen, Eric. 2001. *Games and Information: An Introduction to Game Theory*. 3rd ed. Oxford: Basil Blackwell.

Taylor, T.L. 2006. *Play between Worlds: Exploring Online Game Culture*. Cambridge, MA: MIT Press.

Wark, McKenzie. 2007. *Gamer Theory*. Cambridge, MA: Harvard University Press.

Wikipedia. "Gamification." http://en.wikipedia.org/wiki/Gamification.

Wittgenstein, Ludwig. 1953/1986. *Philosophical Investigations*. Translated by G.E.M. Anscombe. Oxford: Basil Blackwell.

14

Geek

Christina Dunbar-Hester

The *Oxford English Dictionary* defines *geek* as "depreciative. An overly diligent, unsociable student; any unsociable person obsessively devoted to a particular pursuit."[1] This usage goes back to at least the 1950s. The *OED* offers a more recent definition (from the 1980s) of *geek* as "a person who is extremely devoted to and knowledgeable about computers or related technology," and notes that "in this sense, esp. when as a self-designation, not necessarily *depreciative*." Another iteration of *geek* meant a circus freak or carnival performer. This usage is a bit earlier, with the *OED* listing 1919 for a carnival performer and 1935 for the colorful description "a degenerate who bites off the heads of chickens in a gory cannibal show." Precedent for both the circus geek and the academic geek is found in nineteenth-century usage meaning a foolish, offensive, or worthless person.

Geeking as a verb also has a lineage that might surprise us. By 1990, to *geek out* meant to study hard, a denotation linking the phrase to the studious diligence of *geek* as a noun. More specifically, the 1991 *New Hacker's Dictionary* links geeking out to technology, offering the following definition: "*Geek out*, to temporarily enter techno-nerd mode while in a non-hackish context."[2] But in its 1930s incarnation, the *OED* defines *geeking* and *geeking out* as "to give up, to back down; to lose one's nerve. Also with *out*" (in addition to the definition linked to performing as a circus freak). Geeking was originally equated with weakness and failure.

This inversion—from geeking as weakness to geeking as mastery—is worth scrutinizing. Yet even as geeking signifies academic or technical potency, it retains hints of cultural ambivalence. Though geeks may now be celebrated as heroes, they are still characterized in popular culture and the popular press as physically weak, socially maladjusted, and outside of "normalcy" more often than not. Geeks may or may not differ from nerds. On this distinction, science and technology studies scholar Ron Eglash quotes novelist Douglas Coupland, who writes that "a geek is a nerd who knows that he is one."[3] In other words, self-awareness and embrace of one's geeky status are components of geekhood. Geekhood can be borne with pride, whereas nerds are just nerds, dweebs, losers.[4] In popular culture, *geek* can mark outsider status (e.g., the reality television program *Beauty and the Geek*, 2005–8) or outsider status along with studiousness (as in the cult TV show *Freaks and Geeks*, 1999–2000). Katherine Dunn's 1989 novel *Geek Love* centers on carnival geeks negotiating belonging within their family and the wider society.[5]

To explain this drift over time, we might look to the transition from body work to "knowledge work" that has occurred over the twentieth century. As cultural historian Anson Rabinbach explains, throughout the nineteenth century, society was understood to be powered and moved forward by bodies at work: "The human body and the industrial machine were both motors that converted energy into mechanical work."[6] By the late twentieth century, though, the *mind* was ascendant, at least metaphorically. (Industrial and body work still exist, of course, but they have been rendered invisible by a combination of offshore manufacturing in global supply chains and the discursive exaltation of managerial and intellectual work, which marginalizes service work and manual labor.)

Geeks have been caught up in this shift, moving from a position of weakness and marginality to a position of greater relevance and influence. Human minds, not bodies, are understood to be the seat of power in late capitalism. That said, geeks have not moved into unambiguously hegemonic positions, and the geek body retains its status as a site of spectacle. While geeks no longer decapitate chickens with their teeth, they are often portrayed as gawky, puny, and bespectacled[7]—perhaps not monstrous, but still deformed.[8]

The genealogy of *geek* is important for multiple reasons. Not only is it now centered on knowledge (especially arcane knowledge); it has transmuted from a term of insult into a more positive descriptor. Many people use *geek* to describe themselves and others in a fond, self-aware form of teasing and playfulness. As with other iterations of identity politics,[9] geeks have laid claim to a title with a history as a term of disparagement in order to gain power over its use, and they now derive strength from a label that had once been injurious to them. Geeks' embrace of this term now signifies their own uniqueness, their distinctness from the mainstream and commonality with each other.

Notably, *geek*'s acquiring positive valence and in-group signification coincides with computing's rise in prominence over the past three or four decades. Computers have made a leap in the popular imagination from symbols of dehumanizing bureaucracy to intimate machines for self-expression and liberation.[10] Programmers, computing magnates, and hackers have catapulted into the limelight. This is evident in the stature and perceived social power of such figures as Bill Gates, Steve Jobs, and Mark Zuckerberg. Their technical "wizardry"[11] is an object of public reverence. At the same time, especially as geeks shade into hackers, they may be met with suspicion and ambivalence, as Julian Assange and Edward Snowden can attest.[12] This indicates that the freakish, threatening elements of geekhood may still be conjured (see **hacker**). (Programming exhibits a long history of conflict between practitioners and the institutions that employ them, including contestations over requirements for entry and craft versus science status, as computing historian Nathan Ensmenger has shown.)[13]

"Wizardry" is also gendered, of course. Technical masculinity precedes computing.[14] Historian of communication Susan Douglas locates amateur radio operators' work with radio as a site of reinforcement of ideas about masculine identity and technical competence in the early twentieth century.[15] She discusses how the tinkering work performed by men and boys, celebrated in the press, helped attenuate tensions between conflicting definitions of masculinity. Tinkering offered access to a masculine technical domain that was accessible and valued, and that stood in contrast to masculine ideals of ruggedness, strength, and plunder, which were becoming less accessible and less valuable. Douglas's account demonstrates that radio amateurs seized the new technology and interpreted it in a way that emphasized masculinity and different gender roles in relation to it; the technology was used to reinterpret masculinity itself. Electronics tinkering was a remarkably stable elite masculine hobby during the twentieth century, offering suburban men and boys both a masculine space within the domesticity of the home and training for white-collar technical professions.[16] But by the last couple of decades, the object of tinkering had begun to shift away from radio and toward computers.[17]

The continuity of tinkering as a masculine pursuit offers some clues about geek identity. Computer geeks (like the hams before them) are overwhelmingly likely to be white men (or youth), often from middle-class or upper-middle-class backgrounds.[18] Reasons for this likely include exposure to computing at a young age, parental educational achievement, gender expectations and socialization of children and youth, and cultural norms in computer science and hobbyist communities, among others.[19] Geek identity is a factor in the perpetuation of the exclusivity of technical cultures ranging from engineering to Silicon Valley (see **community**, **forum**, and **gaming**).

This is not to say that participation in computing or related technical pursuits is closed to all who are not white men. Strategies to combat the association of geekiness with white masculinity include linking geek identity to technical *engagement* as opposed to technical *virtuosity*.[20] Technical communities including free and open-source software and hackerspaces have repeatedly sought to address issues of "diversity" within their ranks.[21] Women can and do identify as geeks.[22] And Ron Eglash argues that Afro-futurism is an example of an improvised way to achieve technical prowess or identification without being tied to geekiness per se.[23] Yet the association of white middle-class masculinity with aptitude and affection for computing is entrenched.

It is worth locating geeks in space and culture, not only in time. Arguably, to be a geek is to assume a subject position within capitalism, or at least in a technologically advanced society where an abundance of gear and a surfeit of time (whether one's own leisure time/volunteer labor, time stolen from an employer, or something in between)[24] can be presumed.[25] A subsistence farmer is not a geek, no matter how technically adept she is. Not only does *geek* originate within a largely North American or European cultural context, the export of geek identity can be interpreted as a means to bring people in other parts of the world (especially the Global South) into alignment with neoliberal[26] and capitalistic values. It is not a coincidence that some of the values of geek communities, including self-organization and peer production, can be easily ported onto discourses of entrepreneurship and bootstrapping.[27]

Such attempts to export geekhood to, say, "Africans"[28] rightly identify "computer capital" as a "as a mark of distinction with which to ensure their viability on the job and in the social structure."[29] Yet they fail to consider the inadequacy of the "distributive paradigm" as a mode of intervention into systemic inequality.[30] In other words, social power and technical participation are imbricated to such a degree that they may at first glance seem interchangeable, but increasing participation in technology is no guarantee of movement into a more empowered social position.

Despite the towering symbolic value of IT, geek identity as a global subject position faces obstacles. Gender, for example, is constructed and experienced not in isolation but within a matrix of factors that affect social identity, which include class, nationality, ethnicity, and race.[31] Much of what we know about the intersection of gender and technology suffers from the fact that scholars have disproportionately attended to Western cases. Ulf Mellström suggests that we "[need] to investigate configurations of masculinity and femininity in a cross-cultural perspective more thoroughly"[32] in a study illustrating the relative prevalence of women computer scientists in Malaysia. The fact that women are more likely to become computer scientists in Malaysia than in the United States does not necessarily mean it is easier for women to be geeks in Malaysia. Mellström never uses the term at all, and indeed, we would be wrong to conclude that it is an especially meaningful category in this case.[33]

Anthropologist Carla Freeman advocates "localizing" our understandings of work with technology, by which she means attuning any analysis we might conduct to the historical, sociological, geopolitical, and economic factors that materially ground all instances of work with technology.[34] This resonates with the acknowledgment by Wendy Faulkner and many feminist scholars that context matters and "one size does not fit all."[35] Geekhood has the potential to be opened up or modified to fit local conditions of selfhood experienced across nations, genders, or other cultural categories. But every effort should be made to place geekhood, as mode of selfhood and citizenship,[36] within the historical and cultural context from which it emerged. It cannot be held out (or exported) as a universal way of being in the world.

See in this volume: activism, community, democracy, gaming, hacker

See in Williams: capitalism, community, culture, technology, work

Notes

Thanks to Lucas Graves, Lilly Nguyen, Ben Peters, and the Digital Keywords workshop participants for comments on this entry during its development.

1 This obsessive diligence may also be expressed as fandom (Bailey 2005).
2 *Oxford English Dictionary*, online edition, 2014. Emphasis in original.
3 Coupland 1996 quoted in Eglash 2002, n. 1. See also Dunbar-Hester 2008.
4 The *OED* notes that *nerd* has also acquired a definition as a person who pursues a "highly technical interest with obsessive or exclusive dedication." However, it is still more likely to be depreciative, and it is also more broadly defined as "an insignificant, foolish, or socially inept person; a person who is boringly conventional or studious." Much more could be said here. For example, the appearance of the "black nerd" in popular culture indicates that reclamation of *nerd* is possible as well. Significantly, this appropriation (re)codes *nerd* racially, tying African-American-ness to intellectualism. (It thus decouples blackness from primitivism, a linkage exemplified in musician Brian Eno's statement "Do you know what a nerd is? A nerd is a human being without enough Africa in him," quoted in Eglash 2002, 52.) It also expands roles for African Americans beyond "thug, athlete, or rapper." See "The Rise of the Black Nerd in Popular Culture" (*CNN Entertainment*, March 2012), online at http://www.cnn.com/2012/03/31/showbiz/rise-of-black-nerds/. It is worth noting that all the examples of black nerds cited in this piece are men.
5 Thanks to Jack Bratich for discussion of these references, as well as for reminding me that "geeking" can occur around topics other than technology. See also Jason Tocci (2009) on geek identity within popular culture.
6 Rabinbach 1992, 2.
7 See, e.g., Pullin 2009 for a discussion of how prostheses mark disability.
8 Thanks to Ted Striphas and Ben Peters for offering the insight that the circus provides a field for playing out human-nonhuman-animal boundaries, offering a "dirty," transitory space to work out the larger body-mind societal transformation.
9 For example, *queer*. Judith Butler points to a tension for these terms of exclusion, in that even as they are reclaimed and vested with a "positive resignification" (1993, 223), a total metamorphosis, in which past derogatory

valences are cast off, may serve to vitiate their full significance. She cautions that "normalizing the queer would be, after all, its sad finish" (1994, 21). See Dunbar-Hester 2008.

10 Streeter 2011; Turner 2006.
11 Rosenzweig 1998.
12 Gregg and DiSalvo 2013.
13 Ensmenger 2010.
14 And computing was originally women's work. See Abbate 2012; Light 1999.
15 Douglas 1987, chap. 6.
16 Douglas 1987; Haring 2006.
17 See Coleman 2012, 28–30, on youth and coding.
18 See Dunbar-Hester 2008; Kendall 2002; Misa 2010.
19 Kendall 2002; Margolis and Fisher 2003; Misa 2010. See Ensmenger 2010 on the historically tenuous status of programming and the rise of academic computer science.
20 Dunbar-Hester 2008; Dunbar-Hester 2010.
21 Coleman and Dunbar-Hester 2012.
22 Newitz and Anders 2006.
23 Eglash 2002. See also Fouché 2006.
24 Söderberg 2008; Turner 2009.
25 See Coleman 2012; Kelty 2008; Söderberg 2008.
26 Streeter 2011.
27 Streeter 2011, 69–70.
28 See "How Tech Geeks in Africa Are Transforming IT Education" (*Computer World*, April 2012), online at http://www.computerweekly.com/opinion/How-tech-geeks-in-Africa-are-transforming-IT-education.
29 Postigo 2003, 600.
30 See Eubanks 2007.
31 Delgado and Stefancic, scholars of critical race theory, assert that "race and races are products of social thought and relations. Not objective, inherent, or fixed, they correspond to no biological nor genetic reality; rather, races are categories that society invents, manipulates, or retires when convenient" (2001, 7). I invoke race as a category of analysis in light of this insight.
32 Mellström 2009, 886.
33 Mellström argues that two main factors influencing Malaysian women's computer science participation are how Malaysian society constructs appropriate class positions for its multiracial population, and that Malaysian women may embrace "global, corporate masculinity" in part because many Malaysian men reject it for being Western or foreign (2009, 898).
34 Freeman 2000, chap. 3. See also Wyatt 2008.
35 Faulkner 2004, 14.
36 Of course *citizen* is rightly a contentious concept for some. In my use of the term, I wish to signal activity around civic or communal participation, not to marginalize those without full legal status as citizens. Though I do not have space to interrogate "citizenship" here, using it to stand in for a mode of engagement open to "everyone" may present problems.

References

Abbate, Janet. 2012. *Recoding Gender*. Cambridge, MA: MIT Press.
Bailey, Stephen. 2005. *Media Audiences and Identity: Self-Construction and the Fan Experience*. New York: Palgrave Macmillan.
Butler, Judith. 1993. *Bodies That Matter*. New York: Routledge.
———. 1994."Against Proper Objects." *Differences* 6: 1–26.
Coleman, Gabriella. 2012. *Coding Freedom*. Princeton, NJ: Princeton University Press.
Coleman, Gabriella, and Christina Dunbar-Hester. 2012. "Engendering Change? Gender Advocacy in Open Source." *Culture Digitally: Examining Contemporary Cultural Production*, June 26. Online: http://culturedigitally.org/2012/06 /engendering-change-gender-advocacy-in-open-source.
Delgado, Richard, and Jean Stefancic. 2001. *Critical Race Theory*. New York: New York University Press.
Douglas, Susan. 1987. *Inventing American Broadcasting, 1899–1922*. Baltimore: Johns Hopkins University Press.
Dunbar-Hester, Christina. 2008. "Geeks, Meta-Geeks, and Gender Trouble: Activism, Identity, and Low-Power FM Radio." *Social Studies of Science* 38: 201–32.
———. 2010. "Beyond 'Dudecore'? Challenging Gendered and 'Raced' Technologies through Media Activism." *Journal of Broadcasting & Electronic Media* 54: 121–35.
Eglash, Ron. 2002. "Race, Sex, and Nerds: From Black Geeks to Asian American Hipsters." *Social Text* 71: 49–64.
Ensmenger, Nathan. 2010. *The Computer Boys Take Over*. Cambridge, MA: MIT Press.
Eubanks, Virginia. 2007. "Trapped in the Digital Divide: The Distributive Paradigm in Community Informatics." *Journal of Community Informatics* 3(7).
Faulkner, Wendy. 2004. "Strategies of Inclusion: Gender and the Information Society." Final Report (Public Version), SIGIS IST-2000–26329. University of Edinburgh.
Fouché, Rayvon. 2006. "Say It Loud, I'm Black and I'm Proud: African Americans, Artifactual Culture, and Black Vernacular Technological Creativity." *American Quarterly* 58: 639–61.
Freeman, Carla. 2000. *High Tech and High Heels in the Global Economy: Women, Work, and Pink-Collar Identities in the Caribbean*. Durham, NC: Duke University Press.
Gregg, Melissa, and Carl DiSalvo. 2013. "The Trouble with White Hats." *New Inquiry*, November 21. Online: http://thenewinquiry.com/essays/the-trouble -with-white-hats/.
Haring, Kristen. 2006. *Ham Radio's Technical Culture*. Cambridge, MA: MIT Press.
Kelty, Christopher. 2008. *Two Bits: The Cultural Significance of Free Software*. Durham, NC: Duke University Press.
Kendall, Lori. 2002. *Hanging Out in the Virtual Pub*. Berkeley: University of California Press.
Light, Jennifer. 1999. "When Computers Were Women." *Technology & Culture* 40: 455–83.

Margolis, Jane, and Allan Fisher. 2003. *Unlocking the Clubhouse: Women in Computing*. Cambridge, MA: MIT Press.
Mellström, Ulf. 2009. "The Intersection of Gender, Race and Cultural Boundaries, or Why Is Computer Science in Malaysia Dominated by Women?" *Social Studies of Science* (39): 885–907.
Misa, Thomas, ed. 2010. *Recoding Gender: Why Women Are Leaving Computing*. Hoboken, NJ: Wiley-IEEE History Center.
Newitz, Annalee, and Charles Anders, eds. 2006. *She's Such a Geek: Women Write about Science, Technology and Other Nerdy Stuff*. Emeryville, CA: Seal Press.
Postigo, Hector. 2003. "From *Pong* to *Planet Quake*: Post-industrial Transitions from Leisure to Work." *Information, Communication & Society* 6: 593–607.
Pullin, Graham. 2009. *Design Meets Disability*. Cambridge, MA: MIT Press.
Rabinbach, Anson. 1992. *The Human Motor*. Berkeley, CA: University of California Press.
Rosenzweig, Roy. 1998. "Wizards, Bureaucrats, Warriors, and Hackers: Writing the History of the Internet." *American Historical Review* 103: 1530–52.
Söderberg, Johan. 2008. *Hacking Capitalism*. New York: Routledge.
Streeter, Thomas. 2011. *The Net Effect*. New York: New York University Press.
Tocci, Jason. 2009. "Geek Cultures: Media and Identity in the Digital Age." PhD diss., University of Pennsylvania.
Turner, Fred. 2006. *From Counterculture to Cyberculture*. Chicago: University of Chicago Press.
———. 2009. "Burning Man at Google: A Cultural Infrastructure for New Media Production." *New Media & Society* 11: 73–94.
Wyatt, Sally. 2008. "Challenging the Digital Imperative." Inaugural lecture presented upon the acceptance of the Royal Netherlands Academy of Arts and Sciences (KNAW) Extraordinary Chair in Digital Cultures in Development at Maastricht University, March 28. Online: http://www.virtualknowledgestudio.nl/staff/sally-wyatt/inaugural-lecture-28032008.pdf.

15

Hacker

Gabriella Coleman

In the 1950s a small group of MIT-based computer enthusiasts, many of them model train builders/tinkerers, adopted the term *hacker* to differentiate their freewheeling attitude from those of their peers. While most MIT engineers relied on convention to deliver proven results, hackers courted contingency, disregarding norms or rules they thought likely to stifle creative invention. These hackers, like the engineers they distinguished themselves from, were primarily students, but a handful of outsiders, some of them preteens, were also deemed to possess the desire and intellectual chops required to hack and were adopted into the informal club; in the eyes of this group, hackers repurposed tools in the service of beauty and utility, while those students "who insisted on studying for courses" were considered "tools" themselves (Levy 1984, 10).

Since this coinage sixty years ago, the range of activity wedded to the term *hacking* has expanded exponentially. Bloggers share tips about "life hacks" (tricks for managing time or overcoming the challenges of everyday life); corporations, governments, and NGOs host "hackathon" coding sprints (Gregg and DiSalvo 2013; Irani 2015); and the "hacktivist," once a marginal political actor, now stands at the center of geopolitical life (Jordan and Taylor 2004; Beyer 2014; Sauter 2014; Coleman 2014).

Since the early 1980s, the hacker archetype has also become a staple of our mass media diet. Rarely does a day pass without an article detailing an enormous security breach at the hands of shadowy hackers, who have ransacked corporate servers to pilfer personal and lucrative data. Alongside these newspaper headlines, hackers often feature prominently in popular film, magazines, literature, and TV (Alper 2014; Schulte 2013).

Despite this pervasiveness, academic books on the subject of hacking are scant. To date the most substantive historical accounts have been penned by journalists (Levy 1984; Sterling 1992; Lapsley 2013; Greenberg 2012), while academics have written a handful of sociological, anthropological, and philosophical books—typically with a media studies orientation (Thomas 2002; Wark 2004; Kelty 2008; Coleman 2013, 2014). Surveying the popular, journalistic, and academic material on hackers, one discovers that few words in the English language evoke such a bundle of simultaneously negative and positive—even sexy—connotations: mysterious, criminal, impulsive, brilliant, chauvinistic, white knight, digital Robin Hood, young, white, male, politically naive, libertarian, wizardly, entitled, brilliant, skilled, mystical, monastic, creepy, creative, obsessive, methodological, quirky, asocial, pathological.

Some of these associations carry with them a kernel of truth, especially in North America and Europe: conferences are populated by seas of mostly white men; their professionalizable skills, which encompass the distinct technical arts of programming, security research, hardware building, and system/network administration, land them mostly in a middle-class or higher tax bracket (they are among the few professionals who can scramble up corporate ladders without a college degree); and their much-vaunted libertarianism does, indeed, thrive in particular regions like Silicon Valley, the global start-up capital of the world, and select projects like the cryptocurrency Bitcoin.

Yet many other popular and entrenched ideas about hacking are more fable than reality. Hackers, so often tagged as asocial lone wolves, are in fact highly social, as evidenced by the hundreds of hacker or developer cons that typically repeat annually and boast impressive attendance records (Coleman 2010). Another misconception concerns the core political sensibility of the hacker. Many articles universalize a libertarianism to the entirety of hacking practitioners in the West. Whether appraising them positively as freedom fighters or deriding them as naive miscreants, journalists and academics often pin the origins of their practice on an anti-authoritarian distrust of government combined with an ardent support for free market capitalism. This posited libertarianism is most often mentioned in passing as simple fact or marshaled to explain everything from their (supposedly naive) behavior to the nature of their political activity or inactivity (Borsook 2000; Golumbia 2013).

What is the source of this association, and why has it proved so tenacious? The reasons are complex, but we can identify at least two clear contributing factors. First, many hackers, especially in the West, do demonstrate an enthusiastic commitment to anti-authoritarianism and a variety of civil liberties. Most notably, hackers advocate privacy and free speech rights—a propensity erroneously (if perhaps understandably) flattened into a perception of libertarianism. While these sensibilities are wholly compatible and hold affinities with a libertarian agenda, the two are by no means coconstitutive, nor does one necessarily follow from the other.

The second source propping up the myth of the libertarian hacker concerns the framing and uptake of published accounts. Certain depictions of particular aspects of hacking or specific geographic regions wherein libertarianism does, indeed, dominate are routinely represented as, and subsequently taken up as, indicative of the entire hacker culture (Turner 2006).[1] This is only magnified by the fact that Silicon Valley technologists, many of whom promulgate what Richard Barbrook and Andy Cameron have named the "Californian ideology"—"a mix of cybernetics, free market economics, and counter-culture"—are so well resourced that their activities and values, however specific, circulate in the public more pervasively than those at work in other domains of hacker practice (1996). There is no question that the California ideology remains salient (Morozov 2013; Marwick 2013)—but it by no means qualifies as a singular hacker worldview homogeneous across regions, generations, projects, and styles of hacking.

This disproportionately fortified stereotype of the libertarian hacker, along with the paucity of historical studies and contemporary research regarding other values and regional logics at work in hacking, forms the terrain from which scholars of hackers currently work and write. But this seems, slowly, to be changing. Increasingly, scholars are tracing the genealogies of hacking practices, ethics, and values to heterodox, multiplicitous origins (Jordan 2008; Coleman and Golub 2008). For instance, the inception of the "hacker underground"—an archipelago of tight-knit crews who embrace transgression, enact secrecy, and excel in the art of computer intrusion—can be traced to the phone phreaks: proto-hackers who, operating both independently and collectively, made it their mission to covertly explore phone systems for a variety of reasons that rarely involved capital gain (Lapsely 2013). Conversely, "free software" hackers are far more transparent in their constitution and activities as they utilize legal mechanisms that aim to guarantee perpetual access to their creations (Coleman 2012). Meanwhile, "open-source" hackers, close cousins to their equivalents in the free software movement, downplay the language of rights, emphasizing methodological benefits and freedom of choice in how to use software over the perpetual freedom of the software itself; as a result, open-source ideology maintains an affinity with neoliberal logics, while free software runs directly against this current (Berry 2008). Another engagement still is displayed by "the crypto-warriors," covered in great detail by journalist Andy Greenberg, who concern themselves with technical means for securing anonymity and privacy (2012). Their reasons and ideologies differ, but they align in the desire for and development of tools that might ensure these ends.

So while libertarianism is an important worldview to consider, especially in various regions and particular projects, it fails to function effectively as a thread to connect different styles and genres of hacking. However, this doesn't mean we can't consider other commitments around which hackers do, indeed, seem to share a common grounding.

The Craftiness of Craft

Hacking, across its various manifestations, can be seen as a site where craft and craftiness converge: building a 3D printer that can replicate itself; stealing a botnet—an army of zombie computers—to blast a website for a political distributed denial-of-service (DDoS) campaign; inventing a license called copyleft that aims to guarantee openness of distribution by redeploying the logic inherent to copyright itself; showcasing a robot that mixes cocktails at a scientific-geek festival devoted entirely to, well, the art of cocktail robotics; inventing a programming language called Brainfuck which, as you might infer, is designed to humorously mess with people's heads; the list goes on. The alignment of craft and craftiness is perhaps the best location to find a unifying thread that runs throughout the diverse technical and ethical worlds of hacking.

To hack is to seek quality and excellence in technological production. In this regard, all hackers fit the bill as quintessential "craftspeople," as defined by sociologist Richard Sennett: "Craftsmanship names an enduring, basic human impulse, the desire to do a job well for its own sake" (2009). In the twentieth century, with the dominance of Fordist styles of factory labor and other bureaucratic mandates, crafting has suffered a precipitous decline in Western mainstream economies, argues Sennett. Among hackers, however, this style of laboring still runs remarkably deep and strong (Hannemyr 2009).

Even if craftspeople tend to work in solitude, crafting is by definition a collectivist pursuit based on shared rules of engagement and standards for quality. Craftspeople gather in social spaces, like the workshop, to learn, mentor each other, and establish guidelines for exchange and making. Among hackers this ethic has remained intact, in part because they have built the necessary social spaces—mailing lists, code repositories, free software projects, hacker and maker spaces, Internet chat relays—where they can freely associate and work semiautonomously, free from the imperatives and mandates of their day jobs (Shrock 2014).

Large free and open-source projects are even similar to the guilds of yore, where fraternity was cultivated through labor. Free and Open Source Software (F/OSS) institutions are supported by brick-and-mortar infrastructures (servers, code repository) along with sophisticated and elaborate organizational mechanisms. The largest such project is undoubtedly Debian—boasting over a thousand members who maintain the twenty-five thousand pieces of software that together constitute the Linux-based operating system. In existence now for twenty-one years, Debian is a federation sustained by procedures for vetting new members (including tests of their philosophical and legal knowledge regarding free software), intricate voting procedures, and a yearly developer conference that functions as a sort of pilgrimage (Coleman 2013; O'Neil 2009).

Craft and all the social processes entailed—the establishment of rules, norms, pedagogy, traditions, social spaces, and institutions—nevertheless coexist with countervailing, but equally prevalent, dispositions: notably individualism, antiauthoritarism, and craft*iness*. Hackers routinely seek to display their creativity and individuality and are well known for balking at convention and bending (or simply breaking) the rules. If a hacker inherits a code base she dislikes, she is likely to simply reinvent it. One core definition of a hack is a ruthlessly clever and unique prank or technical solution. By extension its creator is also designated as unique.

Craftiness is foremost an aesthetic disposition, finding expression in a plethora of practical engagements that include wily pranks and the writing of code—which is sometimes sparsely elegant and at other times densely obfuscated (Monfort 2008). Its purest manifestation, I have argued elsewhere, lies in the joking and humor so common to the hacker habitat (Coleman 2013, and see the collection in Gorinova 2014). "Easter eggs" provide the classic example: clever and often nonfunctional jokes are commonly integrated into software instructions or manuals.

Hacking is not the only crafting endeavor straddling this line between collectivism and individualism, between tradition and craftiness; the tensions between these poles are apparent among academics who depend upon conventional referencing of peers' work while simultaneously striving to advance clever, novel, counterintuitive arguments and individual recognition. Craftspeople who build and maintain technologies must be similarly enterprising, especially when improvising a fix for something like an old engine or obsolete photocopying machine (Orr 1996). Indeed, the craft-vocation of the security hacker requires what we might describe as intellectual guile. When lecturing to my class one security researcher described the mentality: "You have to, like, have an innate understanding that [a security measure is] arbitrary, it's an arbitrary mechanism that does something that's unnatural and therefore can be circumvented in all likelihood." Craftiness, then, can be seen as thinking outside the box, or circumvention of inherent technological limitations in pursuit of craft. But we can also understand craftiness as exceeding mere instrumentality. Among hackers, the performance of this functional aspect becomes an aesthetic pursuit, a thing valued in and of itself.

The Power and Politics of Hacking

The interplay between craft and craftiness can be treated as something of a hacking universal, then. But it would be wrong to claim that these two attributes are alone capable of sparking political awareness or activism, or even that all hacking qualifies as political, much less politically progressive. Indeed, for a fuller accounting of the politics of hacking it is necessary to consider the variable cultures and ethics of hacking that underwrite craft and craftiness. Hacker political interventions must also be historically situated, in light of regional differences (Chan 2014; Takhteyev 2012), notable "critical events" (Sewell 2005)—like the release of diplomatic cables by the whistle-blowing hacker organization WikiLeaks—and the broader socioeconomic conditions that frame the labor of hacking (Wark 2004).

Indeed, there is little doubt that commercial opportunities fundamentally shape and alter the ethical tenor and political possibilities of hacking. So many hacker sensibilities, projects, and products are motivated by, threatened by, or easily folded into corporate imperatives (Delfanti and Soderberg 2015). Take, for instance, the hacker commitment to autonomy. Technology giant Google, seeking to lure top talent, instituted the "20 percent policy" (Tate 2013). The company affords its engineers, many of whom value technical sovereignty as part of their ethos, the freedom to work one day a week on their own self-directed projects. And Google is not unique; the informal policy is found in a slew of Silicon Valley firms like Twitter, Facebook, Yahoo, and LinkedIn. Of course, critics rightly charge that this so-called freedom simply translates into even longer and more grueling work weeks. Corporations advertise and institutionalize "hackathons" as a way to capitalize on the feel-good mythology of the hacker freedom fighter—all while reaping the fruits of the labor performed therein. In high-tech Chinese cities like Shanghai, where hacker spaces are currently mushrooming, ethics of openness have been determined to bolster entrepreneurial goals beyond those of any individual or unaffiliated collective (Lindtner and Li 2012; Lindtner 2015).

It is nevertheless remarkable that hackers, so deeply entwined in the economy, have managed to preserve pockets of meaningful social autonomy and have frequently instigated or catalyzed political change. Hacking, especially the transgressive art of computer intrusion, to be sure has long exhibited a powerful, albeit latent, political subtext (Soderberg 2012; Wark 2004). But in the past five years, activist-motivated hacking has significantly enlarged its scope and continues to demonstrate nuanced and diverse ideological commitments. Many of these commitments cannot be reduced to "libertarianism," that ideology universalized by many observers as the crux of hacker politics. For one, civil disobedience has surged in a variety of formats and styles, often in relation to leaks and exfiltration. We see lone leakers, like Chelsea Manning, and also collectivist and leftist leaking endeavors, perhaps best exemplified by Xnet in Spain. Other political engagements, similarly irreducible to libertarian values alone, center on collective engagements at the level of software: hackers have recently coded up protocols (like BitTorrent) and technical platforms (like The Pirate Bay) to enable peer-to-peer file sharing and anticopyright piracy (Beyer 2014; McKelvey 2015); since the 1980s, free software hackers have embedded their collectively produced programs with legal stipulations that have powerfully tilted the politics of intellectual property law in favor of access (Kelty 2008; Coleman 2013); Across Europe, Latin America, and the United States, anticapitalist hackers run small but well-functioning collectives that offer privacy-enhancing technical support and services for leftist crusaders; Anonymous, a worldwide protest ensemble specializing in digital direct dissent, has established itself as one of the most populist manifestations of contemporary geek politics—requiring no technical skills to contribute (Coleman 2014); and finally, on the more liberal front, civic and open government hackers throughout North and South America have sought to improve government transparency by creating open standards and applications that facilitate data access and sharing (Gregg and DiSalvo 2013; Schrock forthcoming). Julian Assange, one of the most prominent activist hackers, has recently highlighted the rather dramatic turn to activism and political engagement among geeky technologists. "The political education of apolitical technical people is extraordinary" (2014, 116), he noted during an interview.

If the past five years are any indication, this is a trend that we can expect to grow. What, then, are the sociological and historical conditions that have helped secure and sustain this vibrant sphere of hacker-led political action, especially in light of the economic privilege they enjoy?

Part of the answer lies in craft and the "workshops," like Internet Relay Chat (IRC), mailing lists, and maker spaces, where hackers collectively labor. Taken together they constitute what anthropologist Chris Kelty defines as a recursive public: "a public that is vitally concerned with the material and practical maintenance and modification of the technical, legal, practical, and conceptual means of its own existence as a public; *it is a collective independent of other forms of constituted power and is capable of speaking to existing forms of power through the production of actually existing alternatives*" (2008). What Kelty highlights with his theory of recursive publics is not so much its politics but its *power*—a point also extended in a different manner by McKenzie Wark in *A Hacker Manifesto* (2004). Hackers hold the knowledge—and thus the power—to build and maintain the technological spaces that are partly, or fully, independent from the institutions where they otherwise work for pay. These autonomous zones are where they labor, but also the locales where hacker identities are forged and communities emerge to discuss values deemed essential to the practice of their craft.

Taken from another disciplinary vantage point, these spaces qualify as what sociologists of social movements call "free spaces," historically identified in radical book shops, bars, block clubs, tenant associations, and the like. Generally these are "settings within a community or movement that are removed from the direct control of dominant groups, are voluntarily participated in, and generate the cultural challenge that precedes or accompanies political mobilization" (Polletta 1999). The vibrancy of hacker politics is contingent on the geeky varieties of such free spaces.

It is important to emphasize, however, that while recursive publics or free spaces do not, in and of themselves, guarantee the emergence of hacker political sensibilities, they remain nevertheless vital stage settings for the possibility of activism; however, regional differences figure prominently. For instance, much of the hacker-based political activism emanates from Europe. Compared to their North American counterparts (especially those in the United States), European hackers tend to tout their political commitments in easily recognizable ways, often aligning themselves with politically mandated hacker groups and spaces (Bazzichelli 2013). The continent boasts dozens of autonomous, anticapitalist technology collectives, from Spain to Croatia, and has a developed activist practice that fuses art with hacking (Maxigas 2012). One of the oldest collectives, the German-based Chaos Computer Club (established in 1981), has worked to shape technology policy in dialogue with government for over a decade (Kubitschko 2015). A great majority of the participants populating the insurgent protest ensemble Anonymous are European. Perhaps most tellingly, the first robust, formalized, geek political organization, the Pirate Party, was founded in Sweden (Burkart 2014).

Not all hackers are seeking, however, to promote social transformation. But we can nevertheless consider how many of their legal and technical artifacts catalyze enduring and pervasive political changes regardless of intent.

Craft autonomy figures heavily in this unexpected dynamic, one that can be observed, perhaps most clearly, in the production of F/OSS. Productive autonomy and access to the underlying structures of code are enshrined values in this community, and politics seems to be a natural outcome of such commitments. Irrespective of personal motivation or a project's stated political position, F/OSS has functioned as a sort of icon, a living example from which other actors in fields like law, journalism, and education have made cases for open access. To give but one example, Free Software licensing directly inspired the chartering of the Creative Commons nonprofit, which has developed a suite of open-access licenses for modes of cultural production that extend far beyond the purview of hacking (Coleman and Hill 2004). Additionally, F/OSS practices have enabled radical thinkers and activists to showcase and advocate the vitality, persistence, and possibility of nonalienated labor (Hardt and Negri 2005).

Like F/OSS hackers, those in the underground also strive for and enact craft autonomy with interesting political effects—but here autonomy is understood and enacted differently. Often referred to as blackhats, these hackers pursue forbidden knowledge. Lured as they are by the thrills offered by the subversion of owning and exploring systems, their politics—whether explicit or not—are foremost rooted in transgression for pushing legal, technical, and ethical boundaries. Many of their literary artifacts, such as textfiles and zines, go a step further, actively mocking FBI agents and thus state power (Thomas 2002; Coleman 2012).

Their acts also serve pedagogical purposes, and many have emerged from these illegal, underground nooks into the realm of academic or corporate security research. Their hands-on experiences locating vulnerabilities and sleuthing systems are easily transferrable into efforts to fortify—rather than penetrate—technical systems. Predictably, the establishment of a profitable security industry is seen by some underground hackers as a threat to their autonomy: some critics deride their fellow hackers for selling out to the man (Anonymous 2012). A much larger number don't have a problem with the aim of securitization per se, but nevertheless chastise those attracted to the field by lucrative salaries rather than a passionate allegiance to quality. In one piece declaring the death of the hacker underground, a hacker laments: "Unfortunately, fewer and fewer people are willing, or indeed capable of following this path, of pursuing that ever-unattainable goal of technical perfection. Instead, the current trend is to pursue the lowest common denominator, to do the least amount of work to gain the most fame, respect or money" (Anonymous 2008).

A major, and perhaps unsurprising, motivator of hacker politicization comes in the wake of state intervention. The most potent periods of hacker politicization (at least in the American context) are undoubtedly those following arrests of underground hackers like Craig Neidorf (Sterling 1992) or Kevin Mitnick (Thomas 2002). The criminalization of software can also do the trick; hacker-cryptographer Phil Zimmerman broke numerous munitions and intellectual property laws when he released PGP (Pretty Good Privacy) encryption to the world—a fact governments did not fail to notice or act upon (Levy 2001). But this act of civil disobedience helped engender the now firmly established hacker notion that software deserves free speech protections (Coleman 2009).

In many such instances, the pushback against criminalization spills beyond hacker concerns, engaging questions of civil liberties more generally. Activists outside the hacker discipline are inevitably drawn in, and the political language they deploy results in a sort of positive feedback loop for the hackers initially activated. We saw this precise pattern with the release and attempted suppression of DeCSS, a short program that could be used to circumvent copy and regional access controls on DVDs. In the United States, hackers who shared or published this code were sued under the Digital Millennium Copyright Act, and its author was subsequently arrested in Norway. State criminalization led to a surge of protest activity among hackers across Europe and North America as they insisted upon free speech rights to write and release code, indisputably cementing the association between free speech and code. As alliances were forged with civil liberties groups, lawyers, and librarians, what is now popularly known as the "digital rights movement" was more fully constituted (Postigo 2012).

See in this volume: activism, community, digital, forum, geek, internet, mirror, participation

See in Williams: anarchism, capitalism, collective, creative, culture, democracy, expert, liberation, originality, status

Note

1 For instance, Turner's excellent account about the Silicon Valley regions is taken up to argue for a more general libertarianism.

References

Alper, Meryl. 2014. "'Can Our Kids Hack It with Computers?' Constructing Youth Hackers in Family Computing Magazines." *International Journal of Communication* 8: 673–98.

Anonymous. 2008. "The Underground Myth." *Phrack Inc.* 0x0c, no. 0x41 (November).

———. 2012. "Lines in the Sand: Which Side Are You On in the Hacker Class War." *Phrack Inc.* 0x0e, no. 0x44 (April).

Assange, Julian. 2014. *WikiLeaks*. New York: OR Books.
Barbrook, R., and A. Cameron. 1996. "The California Ideology." *Science as Culture* 6(1): 44–72.
Bazzichelli, Tatiana. 2013. *Networked Disruption: Rethinking Oppositions in Art, Hacktivism and the Business of Social Networking*. Aarhus N, Denmark: Aarhus Universitet Multimedieuddannelsen.
Berry, David. 2008. *Copy, Rip, Burn: The Politics of Copyleft and Open Source*. London: Pluto Press.
Beyer, Jessica L. 2014. *Expect Us: Online Communities and Political Mobilization*. Oxford: Oxford University Press.
Borsook, Paulina. 2000. *Cyberselfish: A Critical Romp through the Terribly Libertarian Culture of High Tech*. New York: PublicAffairs.
Burkart, Patrick. 2014. *Pirate Politics: The New Information Policy Contests*. Cambridge, MA: MIT Press, 2014.
Chan, Anita Say. 2014. *Networking Peripheries: Technological Futures and the Myth of Digital Universalism*. Cambridge, MA: MIT Press.
Coleman, Gabriella. 2009. "Code Is Speech: Legal Tinkering, Expertise, and Protest among Free and Open Source Software Developers." *Cultural Anthropology* 24(3): 420–54.
———. 2010. "The Hacker Conference: A Ritual Condensation and Celebration of a Lifeworld." *Anthropological Quarterly* 83(1): 47–72.
———. 2012. "Phreaks, Hackers, and Trolls and the Politics of Transgression and Spectacle." In *The Social Media Reader*, edited by Michael Mandiberg. New York: New York University Press.
———. 2013. *Coding Freedom: The Ethics and Aesthetics of Hacking*. Princeton, NJ: Princeton University Press.
———. 2014. *Hacker, Hoaxer, Whistleblower, Spy: The Many Faces of Anonymous*. London: Verso.
Coleman, E. Gabriella, and Alex Golub. 2008. "Hacker Practice." *Anthropological Theory* 8(3): 255–77.
Coleman, Gabriella, and Mako Hill. 2004. "How Free Became Open and Everything Else under the Sun." *MC Journal* 7(3) (July).
Delfanti, Alessandro, and Johan Soderberg. 2015. "Hacking Hacked! The Life Cycles of Digital Innovation." *Science, Technology & Human Values* 40: 793–98.
Golumbia, David. 2013. "Cyberlibertarians: Digital Deletion of the Left." *Jacobin*, December 4 https://www.jacobinmag.com/2013/12/cyberlibertarians-digital-deletion-of-the-left/.
Goriunova, Olga, ed. 2014. *Fun and Software: Exploring Pleasure, Paradox and Pain in Computing*. New York: Bloomsbury Academic.
Greenberg, Andy. 2012. *This Machine Kills Secrets: How WikiLeakers, Cypherpunks, and Hacktivists Aim to Free the World's Information*. New York: Dutton Adult.
Gregg, Melissa, and Carl DiSalvo. 2013. "The Trouble with White Hats." *New Inquiry*, November 21. http://thenewinquiry.com/essays/the-trouble-with-white-hats/.
Hannemyr, Gisle. 1999. "Technology and Pleasure: Considering Hacking Constructive." *First Monday* 4(2) (February 1). http://journals.uic.edu/ojs/index.php/fm/article/view/647.

Hardt, Michael, and Antonio Negri. 2005. *Multitude: War and Democracy in the Age of Empire*. Reprint ed. New York: Penguin Books.
Irani, Lili. 2015. "Hackathons and the Making of Entrepreneurial Citizenship." *Science, Technology & Human Values* 40(5): 799–824.
Jordan, Tim. 2008. *Hacking: Digital Media and Technological Determinism*. Cambridge: Polity Press.
Jordan, Tim, and Paul Taylor. 2004. *Hacktivism and Cyberwars: Rebels with a Cause?* Routledge.
Kelty, Christopher M. 2008. *Two Bits: The Cultural Significance of Free Software*. Durham, NC: Duke University Press.
Kubitschko, Sebastian. 2015. "Hackers' Media Practices: Demonstrating and Articulating Expertise as Interlocking Arrangements." *Convergence: The International Journal of Research into New Media* 21(3): 388–402.
Lapsley, Phil. 2013. *Exploding the Phone: The Untold Story of the Teenagers and Outlaws Who Hacked Ma Bell*. New York: Grove Press.
Lavy, Steven. 1984. *Hackers: Heroes of the Computer Revolution*. Garden City, NY: Anchor Press/Doubleday.
———. 2001. *Crypto: How the Code Rebels Beat the Government—Saving Privacy in the Digital Age*. London: Penguin Books.
Lindtner, Silvia. 2015. "Hacking with Chinese Characteristics: The Promises of the Maker Movement against China's Manufacturing Culture." *Science, Technology & Human Values* 40: 854–79.
Lindtner, Silvia, and David Li. 2012. "Created in China." *Interactions* 19(6): 18.
Marwick, Alice E. 2013. *Status Update: Celebrity, Publicity, and Branding in the Social Media Age*. New Haven, CT: Yale University Press.
Maxigas. 2012. "Hacklabs and Hackerspaces—Tracing Two Genealogies." *Journal of Peer Production*, no. 2. http://peerproduction.net/issues/issue-2/peer-reviewed-papers/hacklabs-and-hackerspaces.
McKelvey, Fenwick. 2015. "We Like Copies, Just Don't Let the Others Fool You: The Paradox of The Pirate Bay." *Television and New Media*. 16(8): 734–50.
Montfort, Nick. 2008. "Obfuscated Code." In *Software Studies: A Lexicon*, edited by Matthew Fuller. Cambridge, MA: MIT Press.
Morozov, Evgeny. 2013. *To Save Everything, Click Here: The Folly of Technological Solutionism*. New York: PublicAffairs.
O'Neil, Mathieu. 2009. *Cyberchiefs: Autonomy and Authority in Online Tribes*. New York: Pluto Press.
Orr, Julian E. 1996. *Talking about Machines: An Ethnography of a Modern Job*. Ithaca, NY: ILR Press.
Polletta, Francesca. 1999. "'Free Spaces' in Collective Action." *Theory and Society* 28(1): 1–38.
Postigo, Hector. 2012. *The Digital Rights Movement: The Role of Technology in Subverting Digital Copyright*. Cambridge, MA: MIT Press.
Sauter, Molly. 2014. *The Coming Swarm: DDOS Actions, Hacktivism, and Civil Disobedience on the Internet*. New York: Bloomsbury Academic.
Schrock, Andrew Richard. 2014. "'Education in Disguise': Culture of a Hacker and Maker Space." *InterActions: UCLA Journal of Education and Information Studies* 10(1) (January 1). http://escholarship.org/uc/item/0js1n1qg.

———. Forthcoming. "Civic Hacking as Data Activism and Advocacy: A History from Publicity to Open Government Data." *New Media and Society*.

Schulte, Stephanie Ricker. 2013. *Cached: Decoding the Internet in Global Popular Culture*. New York: New York University Press.

Sennett, Richard. 2009. *The Craftsman*. New Haven, CT: Yale University Press.

Sewell, William H. 2005. *Logics of History: Social Theory and Social Transformation*. Chicago: University of Chicago Press.

Soderberg, Johan. 2012. *Hacking Capitalism: The Free and Open Source Software Movement.* London: Routledge.

Sterling, Bruce. 1992. *The Hacker Crackdown: Law and Disorder on the Electronic Frontier*. New York: Bantam Books.

Takhteyev, Yuri. 2012. *Coding Places: Software Practice in a South American City*. Cambridge, MA: MIT Press.

Tate, Ryan. 2013. "Google Couldn't Kill 20 Percent Time Even If It Wanted To." *Wired*, August 21. http://www.wired.com/2013/08/20-percent-time-will-never-die/.

Thomas, Douglas. 2002. *Hacker Culture*. Minneapolis: University of Minnesota Press.

Turner, Fred. 2006. *From Counterculture to Cyberculture: Stewart Brand, the Whole Earth Network, and the Rise of Digital Utopianism*. Chicago: University of Chicago Press.

Wark, McKenzie. 2004. *A Hacker Manifesto*. Cambridge, MA: Harvard University Press.

16

Information

Bernard Geoghegan

> I think perhaps the word 'information' is causing more trouble in this connection than it is worth, except that it is so difficult to find another that is anywhere near right.
>
> —Claude Shannon

Information as keyword—digital or otherwise—did not exist before the twentieth century. Despite the fact that philological forerunners can be found in ancient Greek and Latin texts and the word *information* appears in some medieval European languages, these terms excited little systematic reflection before the twentieth century. Then, unexpectedly, in the 1920s this formerly unmarked and unremarkable concept became a focal point of widespread scientific and mathematical investigation.[1] In 1948 mathematician Claude E. Shannon of the Bell Telephone Laboratories put forth an enduring account of information in terms of discrete, serial patterns amenable to statistical description, measurable in terms of binary digits (or bits). Information, as the Bell engineer understood, was manageable in terms of coding schemas adapted to the characteristics of the transmitting channel. Over the next few decades this account came to dominate definitions of communication in the sciences and in engineering.[2] Supplementary movements to found schools of information, disciplines of informatics, and to conceptualize the characteristics of an emerging information society followed.[3] This essay examines the changes in natural philosophy, science, and industry that allowed information to emerge as an entity and as a keyword.[4]

Medieval and Early Modern Information

When the word *information* entered Middle English in the fourteenth century, it took its place within a scholastic cosmology wherein resemblance "organized the play of symbols [and] made possible knowledge of things."[5] Derived from Latin *informare*, *information* denoted the imparting of form onto matter. In the earliest extant reference to information in English, dating from 1387, John Trevisa wrote that "fyve bookes com doun from heven for informacioun of mankinde." It nearly misses the point to suggest that medieval information did not "yet" define information in terms of stable, discrete, and serial units for the simple fact that such a conception of medial difference and identity did not yet exist (nor was there any reason to believe such a definition was destined to someday appear). To cite one medium of the period as example, the illuminated medieval manuscript, Jean-François Lyotard has observed that its inscriptions consisted "of differences, which [could not] be transcribed into stable oppositions."[6] Impressions, superimpositions, colors, and a play of lines and figures irreducible to any rational mathematical calculus dominated the page. It would have made no sense to define this play of differences in terms of a stable informational value, be it rational-mathematical or empirical-factual. Information and transmission had more to do with in*spir*ation, or the imparting of intelligible qualities. In this vein the monk William Bonde asserted in 1530 that Christian apostles made their "Crede by instinccyon & informacyon of the holy goost."[7] Repetition and similitude are not the source for information; rather they are the quintessence of its form.

As resemblance lost its grip on the European epistemological imaginary in the course of the sixteenth and seventeenth centuries, an emerging class of natural scientists began to submit traces to schemes of rational and systematic analysis.[8] This technique belonged to a creeping dissatisfaction with explanations of the world in terms of chains of resemblances emanating from divinity. Matter began to take on a brittle and visceral character available for worldly observation and measurement. Eighteenth-century philosopher David Hume signaled how this changing conception shaped the understanding and definition of traces when he identified the term *information* with more or less arbitrary sensory impressions that became intelligible only when submitted to "abstract reasoning or reflection."[9] According to Hume's conception of information, an immaterial form continued to suffuse matter, but the reasoning labors of the human mind—rather than the tracing of spiritual origins—assumed the task of identifying its features. Analogical procedures of in-forming no longer imparted intelligibility; instead, schemes of rational analysis brought order to impressions adrift in empires of empiricist signs. Philosopher Michel Foucault identified Hume's analytical strategy with a broader effort in early modern thought to establish identities through a "means of measurement with a common unit, or, more radically, by its position in an order."[10] The undoing of the great chain of resemblance extended from heaven to earth would ultimately disclose new strings statistically distributed in earthly matter.

Tracing with Telegraphy in the Nineteenth Century

In the nineteenth century technologies of automated inscription took up the analytical slack of exasperated empirical philosophers. Telegraphy—the technique of writing at a distance, often through recourse to electrical signals—took the lead in delineating patterns and series that would eventually be called information. The adaption of telegraphic instruments for inscription and transmission in fields such as physiology, electromagnetism, linguistics, metrology (the science of measurement), spiritualism, and commodity trading generated standardized and discrete traces available for description in technological and mathematical terms.[11] Telegraphic tracing allowed proto-informational measurements of the world.

Three features of telegraphy proved decisive in creating informational entities: instrumentation, graphical standardization, and economization. Classical tools such as a hammer or a fountain pen invested each impression with singular qualities based in part on the human hand that wielded them. By contrast, the telegraphic instrument invested entities with mathematically determined, standard, uniform features. Schemes of mathematical and industrial efficiency organized factors such as the patterning and spacing of letters, and the frequency of distribution among dots and dashes.

Applied to diverse phenomena such as nerve transmissions and railway-switching commands, telegraphy gradually invested a wide range of singular entities with comparable scriptural properties that could be compared to one another.[12] Under telegraphic conditions written language, the actions of the nervous system, and the movements of thc stock market all are reduced to binary discrete notations agreed upon in advance by sender and receiver.[13] Industrial economization provided an imperative for developing a common system of measurement to explain the abstract forms and laws governing the patterning and distributions of these signals. Firms such as the American Telephone and Telegraph Company encouraged engineers to maximize profits by identifying the minimum data and infrastructure necessary to serve customers.[14] In the nineteenth and early twentieth centuries engineers initially referred to the data of transmissions as *intelligence*, swapping out that term for *information* only as it became clear that intelligibility to humans was not necessarily a factor in discerning these patterns.

The Handbook of the Telegraph published in London in 1862 illustrates how telegraphy tended to turn all communications into standardized, quantifiable traces. The guide advises would-be telegraphic clerks that excellent handwriting and basic competency in mathematics (skills associated with creating a standardized and quantified chain of reproduction) will aid them in their quest to become communications professionals. Most remarkable is the one skill it identifies as *nonessential*: the ability to speak or understand the language being telegraphed. "An 'instrument clerk,'" the manual explains, "may be quite competent to telegraph or receive a dispatch in a foreign language and yet not understand a single world of it."[15] What matters is the ability to process discrete letters and patterns with machinelike efficiency and total indifference to the social, cultural, and geographic specificities of clients. "Constant practice," the manual explains, "enables [the telegraph clerk] to signal, *i. e.* to send and receive messages . . . with the rapidity of lightning, hence annihilating distance and concentrating time, conveying tidings of the movements of an army, the rise and fall of dynasties, or the desires of a peasant, with like facility and marvelous speed."[16]

The creation of standardized technical traces operated most evidently at the level of the Morse code, which employed short notations for frequently used letters such as *a*, *e*, and *s* and longer notation for infrequently used letters such as *x* and *z*.[17] Such coding strategies took for granted that signals should be economical and standardized. Typically this meant introducing modes of technical "compression," which communications historian Jonathan Sterne defines as "the process that renders a mode of representation adequate to its infrastructures."[18] This adaptation did not stop at communicated content. Ultimately engineers refashioned transducers, wires, and clerks' bodies as standard equipment for industrialized communications.[19] For example, one study by an engineer at Bell Telephone Laboratories observed that Morse code involved "a tradeoff between code speed and the mean number of hand motions per transmitted letter." Thus even mathematical inefficiencies in Morse code came down to the strategic decision to accommodate the limits of the human channel.[20]

Theorizing Information in the 1920s

The techniques of telegraphy proved more enduring than the electrical telegraph. In the twentieth century, as the economic power of telegraphy waned, the epistemic status of its techniques waxed. Engineers generalized telegraphic methods of instrumentation and economization into a general theory of information applicable across diverse infrastructures.[21] In the 1928 essay "Transmission of Information" Ralph Hartley of the Bell Telephone System proposed substituting for the cognitively connoted term engineers typically applied to transmission patterns, *intelligence*, the less anthropocentric term *information*. Hartley reasoned that human cognition should not feature in the definition of signals. He cited the example of a hand-operated submarine capable of transmitting both messages composed by human beings and those generated by an automatic selecting device. The receiver of such a signal does not assign meaning to the message but only decodes its sequence. Therefore, Hartley posited, "we should ignore the question of interpretation . . . and base our result on the possibility of the receiver's distinguishing the result of selecting any one symbol from that of selecting any other. By this means the psychological factors and their variations are eliminated and it becomes possible to set up a definite quantitative measure of information."[22]

Scrubbing away psychology allowed Hartley to offer a mathematical definition of information applicable to all serially patterned transmissions. He posited that

$$H = n \log s,$$

wherein H designated the quantity of information associated with *n* selections, and *s* stood for the total number of possible selections for a given symbol. This equation defined communication as the unidirectional transmission of serial and discrete messages from a predefined set of symbols. This definition was intuitive for telegraphy, but, Hartley observed, "when we attempt to extend this idea to other forms of communication certain generalizations need to be made."[23] In analyses of media including telephony and television, Hartley showed how communications could be construed as serial representations from a predetermined range of symbolic options. He cited the tendency of these and other forms of electrical communications to render continuous flows from a source as serial, discrete, and predefined coding options. In one of his more peculiar examples of information structures and the relative patterns and freedoms of such selection, Hartley asserted, "In the sentence, 'Apples are red,' the first word eliminates other kinds of fruit and all other objects in general."[24] Thus even spontaneous, ostensibly noncoded and nontechnical communications situations lost their apparently expressive kernel and were replaced by a series of alternating, differential selections. Telegraphy was no longer an informational medium for transmitting speech and meaning; speech and meaning became a medium for the production of telegraphic information.

Information Reformation from 1948

The late 1940s witnessed a groundswell of interest in defining the "stuff" of communications. Widespread and interdisciplinary research into cryptography, computing, radar, and fire control during World War II had stimulated interest in a more precise scientific account of the fundamental laws governing technical communication. Researchers such as Alan Turing, Norbert Wiener, and Claude Shannon worked on diverse communication systems during the war, adapting techniques and conceptions from earlier initiatives to the task at hand, which cultivated an interest in making sense of these common laws. Experience and practice indicated commonalities, but the name or rule of these shared conditions escaped scientific definition. After World War II, *information* emerged as a keyword for defining that commonality. In 1948, at least eight competing accounts of information appeared in prestigious English, British, US-American, and French journals.[25] Claude Shannon of Bell Telephone Laboratories put forth the most influential account. In the opening lines of his 1948 essay "A Mathematical Theory of Communication," published in AT&T's *Bell System Technical Journal*, Shannon asserted that "the fundamental problem of communication is that of reproducing at one point either exactly or approximately a message selected at another point." Shannon made crucial additions to the work of Hartley, including demonstrations of the statistically predictable character of communication signals, showing how redundancy and variable transmission rates could ensure error-free communications, specifying the capacity of communication channels, identifying information with entropy, and postulating binary digits or bits as the most economical measure of transmissions. Throughout the analysis Shannon relied on an analytical framework couched in digital terms (e.g., binary digits) but applicable to analog communications.

When Shannon's theory of communication appeared, it was celebrated but also regarded as a theoretical study of limited practical applicability.[26] His methods of reducing errors through improved coding required expensive digital computers unavailable for general industrial purposes. By the end of the 1950s, engineers widely accepted Shannon as offering the most comprehensive scientific basis for a theory of information, but there was still little expectation of widely implementing the error-correction codes and efficiencies imagined by his analysis. Widespread application appeared only in the 1980s, when falling prices of computers made the development and use of sophisticated digital coding mechanisms economically viable.

The transition from concepts of telegraphic intelligence to *information* as a digital keyword was neither direct nor inevitable. Norbert Wiener, who had received his PhD in philosophy and studied with Edmund Husserl and Bertrand Russell, argued for situating information theories within a grand program of scientific and conceptual synthesis that he termed cybernetics.[27] Wiener proposed a conception of information that was not digital in any essential sense but drew instead on interdisciplinary knowledge to counter scientific specialization.[28] The British physicist Donald MacKay, who worked in radar research during the war, developed a theory of information that drew on Wittgenstein and Calvinist theology for inspiration.[29] Despite these theories' relative resistance to a narrow and technicist conception of information, neither of them really offered a radical critique or alternative to telegraphic reasoning. Both definitions identified information with technical instrumentation, graphical standardization, and economic standardization. In other words, their "alternatives" remained grounded in a communicative cosmology—backlit by the techniques of telegraphy that rested, in turn, upon the analytical strategies of early modern natural philosophy—that had privileged Shannon's methods as the heir apparent to the *information* for the emerging information age.

See in this volume: archive, cloud, culture, digital, flow, sharing

See in Williams: bureaucracy, capitalism, communication, empirical, standards

Notes

Lisa Åkervall and Ben Peters provided incisive commentaries that improved the ideas and phrasing throughout this essay. I thank them for their generous assistance, as well as the Internationales Kolleg für Kulturtechnikforschung und Medienphilosophie for a fellowship that supported the research and writing of this essay.

1 The most instructive and comprehensive accounts of the rise of a scientific definition of information are Friedrich-Wilhelm Hagemeyer, "Die Entstehung von Informationskonzepten in der Nachrichtentechnik" (Free University of Berlin, 1979); William Aspray, "The Scientific Conceptualization of Information," *Annals of the History of Computing* 7(2) (1985): 117–40; Jérôme Segal, *Le Zéro et le Un: Histoire de la notion scientifique d'information*

au 20e siècle (Paris: Editions Syllepse, 2003); John Durham Peters, "Information: Notes toward a Critical History," *Journal of Communication Inquiry* 12(2) (July 1988): 9–23; and Sergio Verdú, "Fifty Years of Shannon Theory," *IEEE Transactions on Information Theory* 44(6) (1998): 2057–78.

2 See Bernard Dionysius Geoghegan, "The Historiographic Conception of Information: A Critical Survey," *IEEE Annals on the History of Computing* 30(1) (2008): 66–81.

3 Ronald Kline discusses intersections between the rise of information theory and schools of information science in "What Is Information Theory a Theory Of? Boundary Work among Scientists in the United States and Britain during the Cold War," in *The History and Heritage of Scientific and Technical Information Systems: Proceedings of the 2002 Conference, Chemical Heritage Foundation*, ed. W. Boyd Rayward and Mary Ellen Bowden (Medford, NJ: Information Today, 2004), 19–24; For an early invocation of the information age, see Marshall McLuhan, *Understanding Media: The Extensions of Man* (Cambridge, MA: MIT Press, 1994), 36. On the characteristics of an information society, see Daniel Bell, "The Social Framework of the Information Society," in *The Computer Age: A 20 Year View*, ed. M. L. Dertoozos and J. Moses (Cambridge, MA: MIT Press, 1979), 500–549.

4 Technical media or *technische Medien* is something of a term of art in Germanophone media studies for communications defined by abstract coding systems that are automated and manipulated with relative autonomy from the human sensorium. Friedrich Kittler identifies the rise of organized, intensively deployed and exploited technical media with the rise of telegraphy. See the bilingual German/English account of this change in Friedrich Kittler, "Geschichte der Kommunikationsmedien," in *Kunst im Netz*, ed. Jörg Huber and Alois Martin (Graz,1993), 73–79.

5 Michel Foucault, *The Order of Things* (New York: Vintage Books, 1973), 17. Tom Gunning discusses the significance of this passage in his forthcoming book, which he presented in part as "Inventing the Moving Image (and Forgetting it Again)," Bauhaus University-Weimar, June 2010. I thank Professor Gunning for informing my argument and analysis here.

6 Cited in Helga Lutz and Bernhard Siegert, "The Ontogenetic Potential of Lines" (unpublished paper).

7 These usages and the definition of *information* come from the *OED* entry on *information*.

8 The most in-depth analysis of this transformation, with a particular concern for the rise of information as a scientific and technological concept, is Bernhard Siegert, *Passage des Digitalen: Zeichenpraktiken der neuzeitlichen Wissenschaften 1500–1900* (Berlin: Brinkmann & Bose, 2003).

9 David Hume, *A Treatise of Human Nature*, ed. L. A. Selby-Bigge (Oxford: Clarendon Press, 1960), 69–70. Emphasis added.

10 Foucault, *The Order of Things*, 55.

11 The most current and comprehensive account of how telegraphy transformed the situation of the sciences, with profound implications for theories of media, is Florian Sprenger, *Medien des Immediaten: Elektrizität, Telegraphie, McLuhan* (Berlin: Kadmos, 2012), 205–330. On electricity, electromagnetism, and telegraphy, with special reference to imperialism

and nineteenth-century commerce, see M. Norton Wise, "Mediating Machines," *Science in Context* 2(1) (1988): 77–113; on metrology and telegraphy, Simon Schaffer, "Late Victorian Metrology and Its Instrumentation: A Manufactory of Ohms," in *Invisible Connections: Instruments, Institutions, and Science*, ed. Robert Bud and Susan Cozzens (Bellingham, WA: Spie Optical Engineering Press, 1992), 23–56; on physiology and telegraphy, see Timothy Lenoir, "Helmholtz and the Materialities of Communication," *Osiris* 9 (1994): 185–207; on spiritualism, electrical experimentation, and telegraphy, see Richard J. Noakes, "Telegraphy Is an Occult Art: Cromwell Fleetwood Varley and the Diffusion of Electricity to the Other World," *British Journal for the History of Science* 32(4) (1999): 421–59; on telegraphy and the changing epistemology of the natural sciences, see Siegert, *Passage des Digitalen*, 334–36 and 359–67.

12 Instruments and instrumentation have been a site of focused research in recent decades. See Don Ihde, *Instrumental Realism: The Interface between Philosophy of Science and Philosophy of Technology* (Bloomington: Indiana University Press, 1991); Albert van Helden and Thomas L. Hankins, eds., *Osiris* 9 (1994); and Thomas L. Hankins and Robert J. Silverman, *Instruments and the Imagination* (Princeton, NJ: Princeton University Press, 1995).

13 On the role of standardization in creating objects for organized scientific inquiry, see Joan H. Fujimura, "Crafting Science: Standardized Packages, Boundary Objects, and 'Translation,'" in *Science as Practice and Culture*, ed. Andrew Pickering (Chicago: University of Chicago Press, 1992), 168–211; and Hans-Jörg Rheinberger, "Scrips and Scribbles," *MLN* 118(3) (April 1, 2003): 622–36. For discussions of how media take part in producing standardized traces, see Mary Ann Doane, *The Emergence of Cinematic Time: Modernity, Contingency, the Archive* (Cambridge, MA: Harvard University Press, 2002), 1–68.

14 Jonathan Sterne, *The Audible Past: The Cultural Origins of Sound Reproduction* (Durham, NC: Duke University of Press, 2002), 32–60.

15 R. Bond, *The Handbook of the Telegraph, Being a Manual of Telegraphy, Telegraph Clerks' Remembrancer, and Guide to Candidates for Employment in the Telegraph Service* (London: Virtue Brothers & Co., 1862), 1.

16 Ibid., 2.

17 In fact, there were a variety of Morse codes and only gradually were they standardized into binary representational schemes. For a discussion of these codes and their relative efficiency, see E. N. Gilbert, "How Good Is Morse Code?" *Information and Control*, no. 14 (1969): 559–65.

18 See Jonathan Sterne, "Compression: A Loose History," in *Signal Traffic: Critical Studies of Media Infrastructures*, ed. Lisa Parks and Nicole Starosielski (Urbana: University of Illinois Press, 2015), 35. For further details on the reconceptualization of language in terms of rational efficiency, see the extraordinary discussion of telegraphy and coding throughout this essay.

19 Kate Maddalena and Jeremy Packer, "The Digital Body: Telegraphy as Discourse Network," *Theory, Culture & Society* 32(1) (January 2015): 93–117.

20 On this point, see Friedrich Kittler, *Gramophone, Film, Typewriter* (Stanford, CA: Stanford University Press, 1999), 2.

21 Thomas Macho notes that such conceptual belatedness is a hallmark of cultural techniques. See Thomas Macho, "Zeit und Zahl: Kalender- und Zeitrechnung als Kulturtechniken," in *Bild, Schrift, Zahl*, ed. Sybille Krämer and Horst Bredekamp (Munich: Wilhelm Fink, 2008), 179.
22 Ralph V. Hartley, "Transmission of Information," *Bell System Technical Journal*, no. 7 (1928): 538.
23 Ibid., 542.
24 Ibid., 536.
25 Verdú, "Fifty Years of Shannon Theory," 2058.
26 See James L Massey, "Deep-Space Communications and Coding: A Marriage Made in Heaven," in *Advanced Methods for Satellite and Deep Space Communications*, ed. Joachim Hagenauer (Heidelberg: Springer-Verlag, 1992), 1–17.
27 See Norbert Wiener, *Cybernetics: Or, Control and Communication in the Animal and the Machine* (Cambridge, MA: MIT Press, 1948); and Norbert Wiener, "The Mathematical Theory of Communication [Review]," *Physics Today* 3 (September 1950): 31–32.
28 Norbert Wiener, "What Is Information Theory?" *I. R. E. Transactions on Information Theory*, June 1956, 48.
29 See Donald M. MacKay, *Information, Mechanism and Meaning* (Cambridge, MA: MIT Press, 1969), 2–3; Donald MacKay, *The Clockwork Image: A Christian Perspective on Science* (Leicester: Inter-Varsity Press, 1974); and Paul Helm, "The Contribution of Donald MacKay," *Evangel* 7(4) (1989): 11–13.

17

Internet
Thomas Streeter

The word *internet* has many referents: hardware, software, protocols, institutional arrangements, practices, and social values. It is not a single thing. But much of its dynamism comes from the fact that, more often than not, the referent is left unspecified; it is spoken about *as if it were* a single object. This has allowed the word to become a kind of metonymy—a part that stands for the whole—for a complex, shifting, intertwined mix of institutions, technologies, and practices. In this it is similar to "the Church," "the press," "Hollywood," or "television." In each case, the use of a part—a building, technology, geographical location, or box in the living room—stands for the whole whatever-it-is.

This metonymic pattern is much more than a convenience. It is an assertion of power. It treats fluid, complex relationships as a self-evident thing, and thereby can cover up instabilities and contested elements within the institutions being considered. This reification, in turn, can help perpetuate or encourage, for better or worse, a specific set of social arrangements. And as those arrangements unfold through time, the metonymy comes to shape the processes it purports to describe. Unpacking *the internet* as a keyword,[1] therefore, offers a window into both the history of the last thirty years and some key political issues of the present.

Someone in a coffee shop might ask the laptop-wielding person next to her, "Are you getting internet?" when she means, "Are you getting a Wi-Fi signal?"—which is actually a local, not internetwork, technology. The term *internet* is often used to refer to a host of different technologies, from non-TCP/IP systems of connection like local area networks and mobile phone data networks, to major "internet backbone" connections involving core routers, fiber-optic long-distance lines, and undersea cables. An "internet connected computer" can mean variously a computer running its own TCP/IP server with its own IP address, or simply any kind of gadget capable of sending some kind of data that will eventually reach global data networks. (A recent news clip referred to "an internet connected umbrella" the handle of which glows when rain is expected, as if "the internet" were the distinguishing technology here rather than, say, the equally essential microchips or wireless technologies.)[2]

The range of multiple meanings go well beyond the technological. A recent headline read, "The 35 Writers Who Run the Literary Internet."[3] This locution assumes that the internet is a separate space or forum apart from other kinds of discussions of literature—even though the community of literary reviewers and their readers actually spans across outlets that vary in terms of both technology (print, digital) and economic organization (profit, nonprofit, advertising supported, subscription, etc.). *Netroots*, a portmanteau of *internet* and *grassroots*, generally refers to progressive left-wing activists who use a mix of traditional and internet forms of political organizing; one does not talk about the Tea Party as a Netroots organization, though it also makes heavy use of the internet and presents itself as grassroots. The internet foregrounded in *netroots* is thus actually a minor part of a politically inflected whole.[4]

Early History: *An* internet vs. *the* Internet

The specific blurriness in how we use *internet* has a history. The root word *network* itself has a history of multiple meanings, in the last century principally divided between an understanding of networks as webs of face-to-face contact without any necessary implication of technological mediation,[5] and networks as technological systems that materially interconnect individuals across distances, such as railroads or telephone systems.[6]

The dual sociological and technological meanings of *network* provided the backdrop when the word *internet* emerged in the 1970s. From the beginning the term expressed some of the tensions and hopes involved in the intertwined problems of technological design and the organization of social relations. *Internetworking* as a verb appeared among computer engineers to mean connecting separate computer networks.[7] This was a response to a social condition, namely, that the first connections of computers across distance occurred in a context largely of private corporations that sold competing systems based on incompatible telecommunications standards. An internetwork was thus something intended to overcome the existing incompatibilities among computer systems from different firms and institutions. Soon shortened further to *internet*, it thus began life as a colloquial term for a particular kind of technological solution to an institutional (rather than purely technical) problem.[8]

A 1977 technical document by Jon Postel, for example, opens,

> This memo suggests an approach to protocols used in internetwork systems. . . . The position taken here is that internetwork communication should be view [*sic*] as having two components: the hop by hop relaying of a message, and the end to end control of the conversation. This leads to a proposal for a hop by hop oriented internet protocol, an end to end oriented host level protocol, and the interface between them. . . . We are screwing up in our design of internet protocols by violating the principle of layering.[9]

In this passage one can see not only the shift from *internetwork* to the shortened *internet*, but also a move from speaking of "networks of networks" in general—"internetwork systems"—toward speaking of the specific system being constructed: "our design of internet protocols." Later in the memo this use of *internet* to refer to a specific system becomes even clearer: "An analogy may be drawn between the internet situation and the ARPANET."[10] In this last passage, "the internet" is clearly being used to refer to the specific system being designed at the time, and thus contrasted with its predecessor network of networks, the ARPANET.

In the next decade, a colloquial use of *internet* to refer to a specific system under construction continued alongside other uses. (And more colloquialisms emerged during this time, such as an even further shortened *the Net*.) During this period, confusion between *internet* as a general principle and as a specific system became of enough concern for engineers of the day to begin to capitalize the latter: *an internet* vs. *the Internet*.[11] But this use of *Internet* to refer to a specific system remained relatively colloquial through the 1980s. At a key moment in 1983, when the existing ARPANET was split into military and research-oriented halves, press reports described the military side as "Milnet" and the civilian side as "R&DNet."[12] While *R&DNet* as a term never caught on, its direct descendant—funded by the National Science Foundation (NSF)—was officially described as NSFNET through the 1980s. In May 1989, the Federal Research Internet Coordinating Committee released a "Program Plan for the National Research and Education Network"; in this instance the committee devoted to internetworking in general uses *Internet* in its self-description, but the proper noun, the specific *thing* being proposed, is called *NREN*.[13]

For the next two years, *Internet* remained an insider's colloquial term for one internetwork among many others, such as Bitnet, BBS systems, Usenet, and the like. As late as December 1992, a famous exchange between vice president–elect Al Gore (who had sponsored the NREN legislation) and the CEO of AT&T, about whether or not government should be involved in the construction of nationwide computer networks, did not contain the word *internet*.[14] The first issue of *Wired*, released the following month, referred to the internet only occasionally in passing, largely as one instance of computer communication systems among others, not as *the* network of networks, not as the center of the "digital revolution" that the magazine was created to celebrate.[15] A May 1993 article in *Newsweek* about the future of computer networks did not mention the internet at all.[16]

The Metonymy Consolidates: 1993–96

All this changed between the fall of 1993 and late 1995, when the contemporary use of *internet* emerged explosively into broad usage, and "the Internet" went from being *an* internetwork to *the* network of networks. By early 1996, the remaining consumer computer communication systems from the 1980s like Compuserve and Prodigy were all selling themselves as means of access to the internet rather than the other way around, the US Congress heavily revised its communications law for the first time in more than half a century in the 1996 Telecommunications Act, major corporations from the phone companies to Microsoft to the television networks were radically revamping core strategies to adapt to the internet, and television ads for Coke and Pepsi routinely displayed URLs.[17] The previously colloquial and unstable term became the fixed name of a global phenomenon.

Though the word became fixed, the phenomenon it referred to was not. For example, the US Telecommunications Act of 1996 was often said to be in part motivated by the rise of the internet, and it referred to "the Internet" several times, defining it rather circularly as "the international computer network of both Federal and non-Federal interoperable packet switched data networks."[18] (X.25 networks, inherited from the 1970s and still in use at the time by banks and other large institutions, were international and packet-switched but were not what the 1996 act was referring to.) The use of "*the* international computer network" instead of "*an* international computer network" thus indicates a referent that was assumed rather than precisely delineated.

What changed in the 1993–96 period was not so much the technology or its reach, but *the way it was imagined*: the shared assumptions, ideas, and values invested in the term took on a new cast and intensity, which in turn shaped collective behavior. It is true that in 1996 there existed a system of material TCP/IP-based computer networking technologies of increasing effectiveness. But the number of nodes and users in that system had been growing logarithmically for several years before 1992 when *internet* was a relatively obscure term, and by the end of 1996 the total number of users remained less than 1 percent of the world population, and less than 8 percent of the US population.[19] By 1996 *the internet* was crystallized as a term, but it was not by any stretch an established central form of communication or means of doing business, and the specific wires and computer systems of which it was made would largely be replaced and transformed within a decade. The material technologies associated with the "internet" therefore were not by themselves as yet all that dominant or settled. The designs, hopes, and money that started flowing toward the thing called the internet in 1996 were based on future expectations, on a shared set of beliefs and visions, as much as on material facts. The "internet" thus was as much a set of ideas and expectations as it was any specific object, yet the habit of referring to it as an object—the metonymy—played a major role in coagulating those ideas and expectations.

Internet as Social Vision: Interactivity, Forum, Telos

So what did the term *internet* refer to, if it did not refer only to an existing technology? One connotation of the term was a particular experience of interactivity: interaction that was widely accessible and designed to be used in an unplanned, playful, or exploratory way, rather than merely as a means to a known end.[20] Most occurrences of *Internet* in the 1996 act are accompanied by the phrase "and other interactive computer services."[21] While not explicitly defined, the "interaction" referred to here was not just any social interaction. In its sociological sense, talking on the telephone is an interaction, a bank official transmitting financial data via an X.25 network is an interaction, but these were already old hat and thus not what was being referred to. The "interaction" in question assumed a certain ease, immediacy, and unplanned type of horizontal connections via connected computers, and wide availability and open access (a fear of which with regard to children was seen in the "Communications Decency" portion of the 1996 act, which forbade pornography on the internet and was subsequently found unconstitutional).

A second important connotation of the internet that emerged was a spatial metaphor, tied to an understanding of it as a kind of *forum*, rather than, say, as a conduit. In the syllabus of the precedent-setting 1997 decision that overturned the "Communications Decency" part of the 1996 act, the US Supreme Court defined the Internet as "an international network of interconnected computers that enables millions of people to communicate with one another in 'cyberspace' and to access vast amounts of information from around the world."[22] (The term *cyberspace* occurs twice more in the decision, without quotes.) Here, to articulate what the internet is, the US Supreme Court casually adopts a spatial metaphor from science fiction (replacing the conduit-oriented "information superhighway" metaphor that dominated in the culture a few years earlier). This spatial metaphor helped ground the court's description of the internet as what it called "this new forum," as a *space* within which citizens interact and deliberate, which thereby underwrote the court's judgment that the internet is worthy of stronger free speech protections than, say, broadcasters.

During this period, the internet also came to be described as having a kind of agency, a force of its own or a teleology. The surprising way in which the internet emerged into broad public consciousness in this period is arguably due to a set of peculiar historical circumstances.[23] But those circumstances were eclipsed by the pressures of the time; it all just seemed to happen as if from nowhere. The resulting shared sense of surprise underwrote a habit of speaking as if it all came from some kind of force attributable to technology alone, without human agency or design. The first (January 1993) issue of *Wired* magazine flamboyantly attributed to the "Digital Revolution" the disruptive force of "a Bengali Typhoon." Over the course of 1993, as the internet came to broad public attention, the magazine began to make such attributions of agency directly to the internet, and thus while not inventing the sense of internet-as-force certainly contributed to its momentum. And this all led to a proliferation of slogans such as "The Net interprets censorship as damage and routes around it,"[24] and a generalized sense that the internet, whatever it was, contained within it a set of inherent traits that had a causal force on society. An entire genre of punditry emerged that exploited the discursive possibilities of this sense of telos: speaking as though one had special insight into the mysterious internet suggested that one had a unique insight into the future, imparting a kind of speaker's benefit similar to that which Foucault pointed out accrues to being an expert on sexuality.[25]

Much of this talk is now easily seen as hyperbolic: it is now routine for various governments to censor their internet, and the late-1990s claim that the internet had somehow suspended the laws of economics encouraged a record-setting stock bubble, with painful consequences when it collapsed. But the sense that the internet has a kind of social force of its own, separable from the intentions and social context of the individuals who construct and use it, persists to this day.

What emerged at the end of the 1993–96 period, in sum, was a meaning of *internet* that unreflectively mixed a shifting set of technologies, protocols, and institutions with connotations of accessible exploratory interaction, a forum, and a sense that the whole "thing" had a teleological causal force. Because this was at a time when the actual systems that we now use were just beginning to be built out, the mix can be seen to have played a constitutive role in those systems' creation.

All these tendencies combine to give the word *internet* an outsize gravitational force in the description of any emerging social practice that has anything at all to do with computer networks. The sense that the internet possesses a kind of agency or telos in particular remains vivid in political and social debates. For example, contemporary net neutrality proponents proclaim, "Save the internet!"—which presumes that the internet, like a National Park or a species of animal, has a kind of natural state of openness, inherent in the internet itself. (As of this writing, it is only recently that less teleological arguments have become more common, such as the argument that net neutrality would help uphold the values of democracy.) An assumption that the internet has a natural telos is also evident in the still-common framing of internet trends as if they represented a natural unfolding rather than economic and social choices. The term *Web 2.0*, borrowing from the tradition of numbered software upgrades, carried with it a sense of an unstoppable progression. A current discourse about a coming "internet of things" similarly implies a kind of "next phase" logic of progression, while implying that the use of wireless and data technologies in home appliances has grand implications, rather than representing merely a continuation of the more than century-long trend in the automation of consumer durables.[26]

Conclusion

It is tempting to draw an analogy, on the level of language, between the emergence of "the internet" in the 1993–96 period and the changes in broad meanings during the World War II period Williams refers to in the opening paragraphs of his original *Keywords*: the experience he had, upon returning from the war, of feeling that "they just don't speak the same language."[27] The war was of course vastly more traumatic and consequential than the appearance of the internet. Yet there is in both cases a sense that in certain circles there were broad shifts in shared meaning, a sense that large groups of people had begun to speak in consequentially new ways. In some ways, the 1993–96 period may be the cultural vortex through which many of the terms discussed in this volume can be said to have either accrued significant new overtones (e.g., **community**, **democracy**, **forum**), emerged into broad usage (e.g., **activism**, **algorithm**, **geek**, **hacker**), or gained new meanings (e.g., **analog**, **cloud**, **memory**, **personalization**). Each term has its own unique history, but the history of the 1993–96 period is essential to understanding most of them.

In the future, the word *internet* might fall into disuse, and historians may wonder why this ill-defined thing called "the internet" received so much attention. From 1990 to 2015, after all, the mobile phone and television both grew dramatically in reach and impact globally, and as of this writing each still has more users than the internet, however defined. Furthermore, eventually platforms like Wikipedia, Facebook, and Netflix may not be framed as things "on the internet," but as the quite distinct institutions they are, with different social, economic, technological, and political implications, as different from one another as, say, cinema and magazines. A future is conceivable in which an internet expert is no more intriguing than a plumbing expert.

Yet future historians will not be able to explain the political, economic, or social histories of the 1990–2015 period without considering the impact of *talk about* "the internet." Laws were passed, stock bubbles inflated and collapsed, political campaigns launched, and a host of influential and broadly shared expectations about politics, economics, and social life were shaped by the term *internet* and the sets of assumptions it carried with it. The word may be vague, but it has mattered nonetheless.

At this point in history, scholars should avoid referring to the internet as a self-evident, single object. But my point here is not simply to denounce the vagueness with which we use the word.[28] We should also pay explicit attention to the hopes, values, and struggles that have been embedded in both the term and the phenomenon. The "internet" may not be the answer, but the questions the term raises are nonetheless crucial. The question of how society designs technologies while also organizing social relations, implicit already in the first casual uses of the term in the 1970s, remains a crucial intellectual and political problem. The "internet" may not be the solution to the problem of democracy, but a democratic future will still need to consider, among other things, questions about technological systems of human interconnection and related political, legal, and economic questions—a problem that has always been embedded in the term itself. And finally, it is significant that one of the great technological triumphs of history was to a significant degree *shaped* by widely shared hopes and visions of democracy and horizontal interactivity, by desires for open forums. The internet may not be inherently democratic, but the fact that we have imagined it thus, that we have invested it with widely shared hopes for democracy, deserves our attention.

See in this volume: activism, algorithm, analog, cloud, community, democracy, forum, geek, hacker, memory, personalization

See in Williams: bureaucracy, community, democracy, ideology, jargon, mechanical, rational, representative, technology

Notes

1 *Internet* is thus a keyword in two senses: the sense that "the problems of its meanings [are] inextricably bound up with the problems it [is] being used to discuss," but also that its meanings are "primarily embedded in actual relationships, . . . within the structures of particular social orders and the processes of social and historical change." Raymond Williams, *Keywords: A Vocabulary of Culture and Society*, rev. ed. (New York: Oxford University Press, 1985), 15, 22.

2 Kristyn Ulanday, "The Internet of Things," *New York Times*, July 16, 2014, http://www.nytimes.com/video/garden/100000003003809/the-internet-of-things.html.

3 http://flavorwire.com/467152/the-35-writers-who-run-the-literary-internet.

4 It has become the norm to speak of information and events as *on* the internet: "I found it on the internet"; "I was arguing with someone on the internet"; "I looked it up on the internet"; "check on the internet." While we say, "I talked to her on the telephone," we would not say, "I found it on the telephone." The telephone is not viewed as its own place so much as a tool to get in touch with specific individuals *across* space. Arguably, one could say the internet is more telephone-like: it is the conduit, whereas individual websites or platforms provide the conditions within which we

are getting information, interacting with others, and so forth. "I saw it on Facebook" or "I looked it up on Wikipedia" are in that sense more accurate. Yet finding or doing something "on the internet" as if it were a location rather than a conduit remains an entirely common way of speaking. The locution for the internet resembles the way we say, "I saw it on television," as opposed to our language for cinema, where we are more likely to say, "I saw it *in* a movie."

5 The tradition is usually said to have begun with Georg Simmel. See, e.g., Mark S. Granovetter, "The Strength of Weak Ties," *American Journal of Sociology* 78(6) (1973): 1360–80.

6 E.g., NBC's Red and Blue radio "networks" of the mid-1920s—though the US 1927 Radio Act and subsequent legal documents referred not to networks but to "chain broadcasting," putting more emphasis on the economic and contractual relationships than on the technological ones.

7 The *OED* finds this use as a verb in 1968: "Internet, v.," *OED Online* (Oxford University Press). The noun form first appears in 1974: "Internet, N.," *OED Online* (Oxford University Press).

8 The continued availability of purely social connotations of *network* and its derivatives, however, is evident in the title of the "Human Rights Internet," appearing in 1981 or earlier, which was a clearinghouse for information about human rights abuses worldwide; to my knowledge it was organized entirely without the use or consideration of computers. See http://www.hri.ca/ or, e.g., David Ziskind, "Labor Laws in the Vortex of Human Rights Protection," *Comparative Labor Law and Policy* 5 (1982): 131.

9 John Postel, "RFC 2.3.3.2 Comments on Internet Protocol and TCP," August 15, 1977, http://www.rfc-editor.org/ien/ien2.txt.

10 Ibid.

11 So, for example, in 1989, an IBM technical manual stated, "When written with a capital 'I,' the Internet refers to the worldwide set of interconnected networks. Hence, the Internet is an internet, but the reverse does not apply." Lydia Parziale et al., "TCP/IP Tutorial and Technical Overview" (IBM, n.d.), 4, http://www.redbooks.ibm.com/redbooks/pdfs/gg243376.pdf. A discussion list created in 1990 to discuss technical and institutional problems with the evolving system was called "Commercialization and Privatization of the Internet" ("com-priv" for short). In this title, the emphasis was already on *the* Internet, not *an* internet. Thomas Streeter, *The Net Effect: Romanticism, Capitalism, and the Internet* (New York: New York University Press, 2011), 110.

12 William J. Broad, "Pentagon Curbing Computer Access; Global Network Split in a Bid to Increase Its Security," *New York Times*, October 5, 1983.

13 Streeter, *The Net Effect*, 107.

14 "The Transition; Excerpts from Clinton's Conference on State of the Economy," *New York Times*, December 15, 1992, New York edition, sec. B.

15 In the "premiere issue," distributed as an iPad-only reissue in 2012, only two out of seven feature articles mention the Internet at all, each case in the sense of a specific system alongside others, such as BBS's, Britain's JANET, and so forth.

16 Jim Impoco, "Technology Titans Sound Off on the Digital Future," *U.S. News and World Report*, May 3, 1993.

17 Streeter, *The Net Effect*, 133–34.

18 Most of the 106-page 1996 act addresses well-established telecommunications systems, e.g., "the general duties of telecommunications carriers." Federal Communications Commission et al., "Telecommunications Act of 1996," *Public Law* 104(104) (1996): 84.

19 http://www.internetworldstats.com/emarketing.htm; Farhad Manjoo, "Jurassic Web," *Slate*, February 24, 2009, http://www.slate.com/articles/technology/technology/2009/02/jurassic_web.html.

20 The once-common term *information retrieval* captures the opposing sense of the use of online communication for a preplanned purpose.

21 E.g., "The rapidly developing array of Internet and other interactive computer services available to individual Americans represent an extraordinary advance." Federal Communications Commission et al., "Telecommunications Act of 1996," 83.

22 Reno, Attorney General of the United States, et al. v. American Civil, U.S. (U.S. Supreme Court 1997).

23 Streeter, *The Net Effect*, 119–37.

24 https://en.wikiquote.org/wiki/John_Gilmore.

25 One might rewrite what Foucault said about the repressive hypothesis by replacing references to sexuality with "internet revolution," thus:

> There may be another reason that makes it so gratifying for us to define the relationship [between technology and society in terms of revolution]: something that one might call the speaker's benefit. [If the internet is revolutionary], then the mere fact that one is speaking about it has the appearance of a deliberate transgression. A person who holds forth in such language places himself to a certain extent outside the reach of power; he upsets established law; he somehow anticipates the coming freedom. . . . [when we speak about the internet] we are conscious of defying established power, our tone of voice shows that we know we are being subversive, and we ardently conjure away the present and appeal to the future, whose day will be hastened by the contribution we believe we are making. Something that smacks of revolt, of promised freedom, of the coming age of a different law, slips easily into this discourse." (Michel Foucault, *The History of Sexuality: An Introduction* [New York: Knopf Doubleday Publishing Group, 2012], 6–7)

26 http://en.wikipedia.org/wiki/Internet_of_Things.

27 Williams, *Keywords*, 11.

28 For a lively example of denunciation, see Evgeny Morozov, *To Save Everything, Click Here: The Folly of Technological Solutionism* (New York: PublicAffairs, 2013), 21.

References

Broad, William J. "Pentagon Curbing Computer Access; Global Network Split in a Bid to Increase Its Security." *New York Times*, October 5, 1983.

Federal Communications Commission et al. "Telecommunications Act of 1996." *Public Law* 104 (104) (1996): 1–5.
Foucault, Michel. *The History of Sexuality: An Introduction*. New York: Knopf Doubleday Publishing Group, 2012.
Granovetter, Mark S. "The Strength of Weak Ties." *American Journal of Sociology* 78(6) (1973): 1360–80.
Impoco, Jim. "Technology Titans Sound Off on the Digital Future." *U.S. News and World Report*, May 3, 1993.
Manjoo, Farhad. "Jurassic Web." *Slate*, February 24, 2009. http://www.slate.com/articles/technology/technology/2009/02/jurassic_web.html.
Morozov, Evgeny. *To Save Everything, Click Here: The Folly of Technological Solutionism*. New York: PublicAffairs, 2013.
Parziale, Lydia, et al. "TCP/IP Tutorial and Technical Overview." IBM, n.d. http://www.redbooks.ibm.com/redbooks/pdfs/gg243376.pdf.
Postel, John. "RFC 2.3.3.2 Comments on Internet Protocol and TCP," August 15, 1977. http://www.rfc-editor.org/ien/ien2.txt.
Streeter, Thomas. *The Net Effect: Romanticism, Capitalism, and the Internet*. New York: New York University Press, 2011.
"The Transition; Excerpts from Clinton's Conference on State of the Economy." *New York Times*, December 15, 1992, New York edition, sec. B.
Ulanday, Kristyn. "The Internet of Things." *New York Times*, July 16, 2014. http://www.nytimes.com/video/garden/100000003003809/the-internet-of-things.html.
Williams, Raymond. *Keywords: A Vocabulary of Culture and Society*. Rev. ed. New York: Oxford University Press, 1985.
Ziskind, David. "Labor Laws in the Vortex of Human Rights Protection." *Comparative Labor Law and Policy* 5 (1982): 131.

18

Meme

Limor Shifman

Once upon a time there was a term. Its name was *meme*. The proud father, Richard Dawkins, showered attention on it while others ridiculed and dismissed it. Then along came the internet, whose users crowned the meme with such popularity, the term almost ran away with them.

The following essay unpacks this simplistic and suggestive fairytale. While *Meme* remains a troublemaker among other titanic digital keywords (see **community**, **culture**, **digital**, **internet**), it is a concept worth deciphering. This term encapsulates more than popular fads such as Gangnam Style, Lolcats, or the ALS Ice Bucket Challenge; it is a crucial conceptual and empirical tool for understanding how digital culture works. Rehearsing Raymond Williams's notion that keywords capture *activities and their interpretation* as well as certain *forms of thought*, I claim that *memes*—or, more precisely, their recent incarnation as *internet* memes—are pivotal for understanding both digital behaviors and the cultural logics governing them.

Memes in the Predigital Era

The term *meme*, unlike many other keywords, can be given a fairly precise birthdate: Richard Dawkins coined it in chapter 11 of his 1976 book *The Selfish Gene*. Whereas the book as a whole is about evolutionary biology, chapter 11 is dedicated to human culture (and its analogy to the world of gene-based replication). Dawkins observes that humans have devised small *cultural* units of transmission that, like genes, spread from person to person through imitation. These units—which evolve and propagate much faster than genes do—follow the same basic principles of variation, competition, selection, and retention; examples include texts (e.g., "Happy Birthday" or "Humpty Dumpty"), ideas (e.g., evolution or heaven), and behaviors (e.g., the miniskirt fashion or blowing out candles on a birthday cake). Dawkins called these units *memes*—a term he derives by shortening the Greek root *mimeme* ("something which is imitated") to nearly rhyme with *gene*. In addition, he draws out the affinity between *meme* and the French word *même* ("same"), as well as its resemblance to *memory* as a source of inspiration (see **memory**). Interestingly, the Greek root *mneme* ("memory") was already being used in the nineteenth century by Austrian sociologist Ewald Hering to discuss cultural evolution. It was further developed by German biologist Richard Semonasa and even adopted as the title of his 1904 book, *Die Mneme*. When he coined the term, Dawkins was unaware of these earlier uses; his fresh appeal to an ancient root came to practice what it preached: it too became a successful meme.

But it took some time. The emergence of memetics—"the theoretical and empirical science that studies the replication, spread and evolution of memes" (Heylighen and Chielens 2009)—was slow and not without obstacles. Its spread took place in part thanks to the contribution of prominent scholars Douglas Hofstadter (1985) and Daniel C. Dennett (1990, 1995), a short-lived *Journal of Memetics* (1997–2005), and several related books around the turn of the millennium (Blackmore 1999; Distin 2005; Lynch 1996). These "meme enthusiasts" soon met with fierce opposition, however. A main criticism related to the concept's ambiguity: even today, debate rages over what precisely a meme is. According to Dawkins's initial definition, memes can be ideas, practices, or texts. However, this broad scope complicates attempts at quantifying and tracking memes' propagation: few can agree on what exactly should be measured. Others have criticized the analogy between culture and nature, and between human behavior and genes in specific, as reductive, materialistic, and insufficient for describing complex human behaviors. A third point of criticism related to the notion that humans are "controlled" by memes. This idea, foregrounded in prominent works such as Blackmore's best seller, *The Meme Machine*, generated resentment and debate over the diminished role of human agency in memetic diffusion. Finally, as the millennium came to a close, some critics claimed that memetics, or the study of memes, has no added value: it does not offer tools or insights beyond anything already in use by cultural anthropologists and linguists (Rose 1998; Chesterman 2005).

Enter: The Internet

While researchers continue arguing about the usefulness of this construct, netizens have delivered their verdict. By the end of the first decade of the twenty-first century, the term *meme* had become an integral part of online vernacular. The concept is now used casually and ubiquitously, mostly to signal the rapid propagation of images, videos, and catchphrases on the internet (Knobel and Lankshear 2007). Within a short period, two significant changes occurred: *meme* often signified not a general unit of culture but a *digital* one, and it moved from the realm of academia (and popular science) to the public sphere (and popular culture). As its own evolutionary logic suggests that a meme succeeds when certain social, cultural, psychological, and technological conditions expedite its uptake, we need ask which factors facilitated memes' memetic success in the third millennium. In what follows, I highlight four such factors—(1) scale, (2) transformation, (3) transparency, and (4) structure. Each of the factors relates not only to technological affordances, but also to the nesting of these affordences in "meme nurtering" social and cultural norms.

Scale. Dawkins's initial framing highlighted three qualities that enhanced memes' success: longevity (survival over time), fecundity (the number of copies produced within a time unit), and copying fidelity (accuracy). As early commentators noted (Marshall 1998), the internet has boosted all three: digital video meme transmission has much higher copying fidelity than oral or print communication; the internet facilitates the rapid diffusion of any given message to numerous recipients; and longevity can also potentially be enhanced, owing to augmented storage possibilities. Put simply, the internet affords broad, quick, and accurate meme propagation.

Transformation. Digital media transform content fairly easily. They offer tools and technologies for imitation and remixing media content. Instead of memorizing long oral epic poems, users are mimicking and remixing digital content, Photoshopping images, or layering in sound tracks, with a few finger clicks. Internet memes evolve and mutate with profound ease thanks to the transformative power of digital tools. As detailed below, this expedited mutation rate may require an amended definition of memes, which will take into account their growing variability in the digital age.

Transparency. Memes not only propagate more quickly and with greater variation on the internet; their transformation may also be, under certain conditions, more transparent. Most popular platforms and applications let lay users access and publicly display metadata about viewing preferences, choices, and user responses in the aggregate. Visible metadata—the number of views, comments, or likes—in turn becomes part of the ratcheting-up process of memetic propagation, where the popular memes become more popular. At the same time, many mechanisms of meme transmission are concealed, leading to powerful modes for hidden network gatekeeping and user surveillance (Nahon and Hemsley 2013).

Structure. Finally, the internet in general, and Web 2.0 in particular, embodies a structural logic congruent with meme diffusion. While memes are passed along from person to person, they gradually scale to constitute shared social phenomena. They spread on a micro basis, but impact the mezzo and sometimes even the macro levels of societies, shaping mind-sets and actions. In the so-called Web 2.0, or participatory culture era, we see the convergence of memes with the notion that bottom-up user-generated modes of transmission have transformative power. Applications such as Facebook, Twitter, and YouTube not only facilitate the diffusion of such content but also incorporate a normative imperative that values "sharing" (John 2013; see also **sharing**). Sharing content—or spreading memes—has recently become a governing logic, part and parcel of the participatory zeitgeist. To paraphrase Ethan Zuckerman (2008), the internet is not made of cats; it is made of memes.

From Memes to *Internet* Memes

So far, I have described memes' compatibility with digital culture. Yet the internet has also dramatically altered memes' modes of circulation. In a recent talk, while noting the broad continuities between viral content and memes offline and online, Richard

Dawkins claimed that internet users have "hijacked" his original idea: while according to his theory memes mutated by random change, now they are altered deliberately by creative individuals (and resourceful marketers in search of hits).[1] The result, whether we are critical of it or not, allows us to reconceptualize the active role individuals play in creating, spreading, and experiencing memes; It also requires an updated definition of the concept, or, to be more precise, it calls for the *Internet* meme to be defined.

Early accounts of internet memes tended to associate them with humor. For example, Christian Bauckhage (2011) claimed that memes "are inside jokes or pieces of hip underground knowledge that many people are in on"; and Patrick Davison (2012) noted that "an Internet meme is a piece of culture, typically a joke, which gains influence through online transmission." Yet as the term appeared more frequently in public spheres—and as it was invoked as part of political and activist agendas (Bennett and Segerberg 2012; Milner 2013)—its automatic association with humorous communication diminished, opening the way for alternative definitions.

Elsewhere I have suggested that the definition of internet memes should entail a shift from singular to plural. Instead of depicting the meme as a *single* cultural unit that has propagated vigorously, internet memes should be defined as *groups* of content units, or as textual families (Shifman 2013). This shift derives from the new ways in which memes are experienced in the digital age. In the past, individuals were likely to be exposed to one meme version at a given time (for instance, they heard a nursery rhyme and only much later encountered a different version of it). In contrast, in digital environments people experience memes as boundless groups of interconnected texts, images, and videos, separated by nothing more than mouse clicks. Increasingly, people are exposed to memetic multiples and *variation*, as successful internet memes tend to incorporate numerous visible versions.

Accordingly I suggest defining an internet meme as follows:

> (a) a group of digital items sharing common characteristics of content, form, or stance, which (b) were created with awareness of each other, and (c) were circulated, imitated, and transformed via the internet by many users. (Shifman 2013)

More than a transition from singular to plural, this definition also incorporates the idea that internet memes may entail several *memetic dimensions*—namely, several aspects that people may imitate. In some memes the binding quality may be a certain type of content (the idea of a cute cat); in others, the imitated element will be a format (a photo with a caption) or a specific stance or position taken by its users (playful, humorous communication). Many memes entail a combination of at least two of these three memetic dimensions.

This suggested definition also highlights the main feature differentiating internet memes from their close relatives, "viral" content. If memes are based on variability between texts, the term *viral* mostly signifies the quick and vast propagation of a *specific* text, video, or image (Nahon and Hemsley 2013). In other words: while we can think of the viral as a singularity that is circulated without significant alterations (for example, the clip Gangnam Style), memes are all about mutation—people's creative reactions to former versions (in this case: the vast body of reactions and remakes that the Gangnam Style video provoked). However, the distinction between memes and virals is fuzzy. "Purely" viral content probably does not exist, since once a photo, or a video, becomes sufficiently popular online, some user, somewhere, will surely do something with it. Similarly, purely memetic content, no matter how unpopular, must involve some degree of cultural contagion or spreading to be considered memetic. Responding to the growing use of *memes* and *virals*, Henry Jenkins and his colleagues (2013) suggest jettisoning both terms and their biological baggage for what they call "spreadable media." While the clarity and sensibility of this concept is appealing, I contend that the rootedness of both memes and virality in contemporary digital discourse requires careful rearticulation rather than replacement of these keywords.

Internet Memes between Ritual and Transmission

Memes have come full circle, from academia to the popular domain and back. In the last decade, internet memes have begun feeding an expanding field of study (Burgess 2008; Gal, Shifman, and Kampf 2015; Knobel and Lankshear 2007; Milner 2012; Miltner 2014; Segev et al. 2015).[2] By way of conclusion, I wish to highlight the ways in which meme studies may employ and bridge James Carey's (1989) classic differentiation between *transmission* and *ritual* models of communication. "Transmission" likens the spread of information through media to the movement of peoples or goods, and asserts that communication is mainly about increasing messages' spread and effect as they travel in space. Contrarily, the "ritual" model looks at communication as a *shared* action—a choreographed construction of communities through shared symbols and practice.

Both approaches seem highly relevant to our understanding of internet memes. In fact, it is impossible to understand memes without amalgamating the two. Since memetic diffusion is based on gradual propagation, the transmission and movement of memes is obviously essential; at the same time, this constant and seemingly chaotic movement of items has a larger social purpose, which often includes community building and reparation (Gal, Shifman, and Kampf 2015; Milner 2012; Miltner 2014). Such meme-based communities are vivid illustrations of the "networked self" (Papacharissi 2010) who, on the one hand, strives to express individuality and creativity (through the articulation of his/her *own* unique meme version) and, on the other, longs for communality (constructed through the shared referencing of the recurring unit). This intersection of the transmission and the ritual models—which requires the simulations investigation of *how* and *why* memes propagate and mutate—outlines a worthy agenda for internet memes studies likely to endure for years to come.

See in this volume: community, culture, digital, internet, memory, sharing

See in Williams: aesthetic, art, behavior, collective, creative, ecology, evolution, generation, genetic, image, mechanical, nature, organic, popular, sociology, structure

Notes

1 Available at http://www.webcitation.org/6HzDGE9Go.
2 Yet memes continue to be an integral part of internet users' mundane vocabulary. A comparison between the academic database Web of Science and Google Trends reveals a striking correspondence between scientific

and popular accounts of the term *meme*, characterized by a sharp rise in the concept's use since 2010.

References

Bauckhage, C. 2011. "Insights into Internet Memes." Paper presented at the 5th International AAAI Conference on Weblogs and Social Media, July, Barcelona, Spain.

Bennett, L., and A. Segerberg, A. 2012. "The Logic of Connective Action." *Information, Communication & Society* 15(5): 739–68.

Blackmore, S. 1999. *The Meme Machine*. Oxford: Oxford University Press.

Burgess, J. 2008. "All Your Chocolate Rain Are Belong to Us? Viral Video, YouTube and the Dynamics of Participatory Culture." In *Video Vortex Reader: Responses to YouTube*, edited by G. Lovink and S. Niederer, 101–10. Amsterdam: Institute of Network Cultures.

Carey, J. W. 1989. *Communication as Culture: Essays on Media and Society*. Boston: Unwin-Hyman.

Chesterman, A. 2005. "The Memetics of Knowledge." In *Knowledge Systems and Translation*, edited by H. V. Engberg and H. Gerzymisch-Arbogast, 17–30). Berlin: Mouton de Gruyter.

Davison, P. 2012. "The Language of Internet Memes." In *The Social Media Reader*, edited by Michael Mandiberg, 120–36. New York: New York University Press.

Dawkins, R. 1976. *The Selfish Gene*. Oxford: Oxford University Press.

Dennett, D. C. 1990. "Memes and the Exploitation of the Imagination." *Journal of Aesthetics and Art Criticism* 48: 127–13. https://ase.tufts.edu/cogstud/dennett/papers/memeimag.htm.

———. 1995. *Darwin's Dangerous Idea: Evolution and the Meanings of Life.* New York: Touchstone.

Distin, K. 2005. *The Selfish Meme: A Critical Reassessment.* Cambridge: Cambridge University Press.

Gal, N., L. Shifman, and Z. Kampf. 2015. "It Gets Better: Internet Memes and the Construction of Collective Identity." *New Media & Society*: 1461444814568784.

Heylighen, F., and K. Chielens. 2009. "Cultural Evolution and Memetics." In *Encyclopedia of Complexity and System Science*, edited by B. Meyers. http://pespmc1.vub.ac.be/Papers/Memetics-Springer.pdf.

Hofstadter, D. R. 1985. *Metamagical Themas: Questing for the Essence of Mind and Pattern*. New York: Basic Books.

Jenkins, H., S. Ford, and J. Green. 2013. *Spreadable Media: Creating Value and Meaning in a Networked Culture*. New York: New York University Press.

John, N. A. 2013. "Sharing and Web 2.0: The Emergence of a Keyword." *New Media & Society* 15(2): 167–82.

Knobel, M., and C. Lankshear, C. 2007. "Online Memes, Affinities and Cultural Production." In *A New Literacies Sampler*, edited by M. Knobel and C. Lankshear, 199–227. New York: Peter Lang.

Lynch, A. 1996. *Thought Contagion: How Belief Spreads through Society*. New York: Basic Books.

Marshall, G. 1998. "The Internet and Memetics." Presented at 15th International Congress on *Cybernetics Symposium on Memetics*, August 24–28, Namur, Belgium. http://pespmc1.vub.ac.be/conf/memepap/marshall.html.
Milner, R. 2012. "The World Made Meme: Discourse and Identity in Participatory Media." PhD diss., University of Kansas.
———. 2013. "Pop Polyvocality: Internet Memes, Public Participation, and the Occupy Wall Street Movement." *International Journal of Communication* 7: 2357–90.
Miltner, K. M. 2014. "'There's no place for lulz on LOLCats': The Role of Genre, Gender, and Group Identity in the Interpretation and Enjoyment of an Internet Meme." *First Monday* 19(8).
Nahon, K., and J. Hemsley. 2013. *Going Viral*. Cambridge: Polity Press.
Papacharissi, Z., ed. 2010. *A Networked Self: Identity, Community, and Culture on Social Network Sites*. London: Routledge.
Rose, N. 1998. "Controversies in Meme Theory." *Journal of Memetics—Evolutionary Models of Information Transmission* 2. http://cfpm.org/jom-emit/1998/vol2/rose_n.html.
Segev, E., A. Nissenbaum, N. Stolero, and L. Shifman. 2015. "Families and Networks of Internet Memes: The Relationship between Cohesiveness, Uniqueness, and Quiddity Concreteness." *Journal of Computer Mediated Communication* 20: 417–33.
Shifman, L. 2013. *Memes in Digital Culture*. Cambridge, MA: MIT Press.
Zuckerman, E. 2008. "The Cute Cat Theory Talk at Etech." March, 8. http://www.ethanzuckerman.com/blog/2008/03/08/the-cute-cat-theory-talk-at-etech/.

19

Memory

Steven Schrag

> If men learn this, it will implant forgetfulness in their souls; they will cease to exercise memory because they rely on that which is written, calling things to remembrance no longer from within themselves, but by means of external marks What you have discovered is a recipe not for memory, but for reminder.
>
> —Plato, *The Phaedrus*

> Reminders. Now nothing slips your mind.
>
> —Apple Inc., "OS X Apps"

Memory—from the Latin *memoria* (the faculty of remembering, remembrance, a historical account), and "mnemonic" from Mnemosyne, mother of the nine Muses in Greek myth (see also **meme**)—is one of the most fundamental concepts of human identity, as well as one of its oldest technologies. It is a process of narrating and making sense of experience, of storage and recovery, at both individual and collective levels. While computer "memory" (itself an expansive category of devices used to store and recall data or programs) is only one mnemonic technology among many, increasingly ubiquitous digital data storage has had a profound effect on contemporary practices of history and remembrance—and even on the way humans construct and perceive their identities. Discussions of a "modernity that forgets" or an "Internet that remembers" often risk conflating individual embodied memory, collective and cultural memory, historical records, storage media, and the archive (see also **archive**). The ways these categories intersect, conflict, and translate across one another have long served as sites of memory power and politics.

Memory is a polysemic term whose uses rest on the tensions between dueling categories in continual contestation. It, like the term *identity*, comprises the particular and the universal, the natural and the artificial, the individual and the collective, the internal and the external. Historically construed as an art, practiced as a technique in oral societies, retained in objects and sites, today memory both mediates and is mediated by new analogies between the brain and the digital records of the hard drive. Drawing on diverse sources of our modern imagination of memory, including science fiction, modern industry, and Western thought, this essay outlines several of these tensions—the paradoxical relation between remembrance and history, the apparent tension between technologically induced amnesia and hypermnesia, and the emergent gaps between mnemonic persistence and ephemerality that shape social structures of domination and control, among others—and address how digital memory becomes political in its intersections with individual and cultural memory.

The Paradox of Prosthesis

Conversations about the relationship between technology and memory stretch back to Plato's injunction in the *Phaedrus*, in which he argued that the act of creating an archive (an argument committed to paper, or today, a file in a cabinet or an email on a server) is a fundamental act of forgetting. To "externalize memory" to storage media is to no longer commit it to cognitive memory. External records also reconstruct memory—not only extending our cognitive memories across time and space, but manipulating them as media prostheses have long done.

Representational practices thus occupy an uneasy place within our understandings of memory. Derrida pathologizes the psychic process of archiving as an endless reiteration akin to the death drive, which he calls "archive fever": it marks the body violently, leaving only an imperfect trace of the original (1995). Memory studies, with strong roots in trauma theory and Freudian psychoanalysis, considers how memory can bridge violent ruptures between the present and the past. According to Caruth, the aporia, or internal contradiction, of trauma is the disjunction between the event and its understanding: "a history can only be grasped in the very inaccessibility of its occurrence" (1996; see also **event**). Theories of socially mediated trauma, in which collective identities confront violent histories, underscore the fact that the mnemonic is also memetic: the act of commemoration, whether as a ritual process of storytelling or an intrusive flashback, indexically "relives" an inaccessible past. Memory is not just the stored past: it is an active process of retrieval, rearrangement, and revival that, like an archive, calls to mind that which it archives. This is not simply to acknowledge the obvious shakiness and suggestibility of human memory: it is rather to suggest that every act of remembering is unsettled by attempts to bind the past to the present, and to orient the present to a revisionist sense of the future.

Two parallel paradigms propose that memory media threaten to turn us all into science fiction prostheses of ourselves—understood either as a liberating cybernetic extension of self or as a dangerous, disembodied Other. The cybernetic imaginary, a fusion of bodies and machine memory, suggests a man-machine communication control system, perhaps gesturing toward a transcendent, utopian global consciousness. By contrast, the postmodern cyberpunk imaginary, a disembodiment of what was once essentially ours—our faculties of recall—into a distant other, points to a bleaker future in which memory media speed us toward a traumatic decline in human integrity and experience (Luckhurst 2008).

Both imaginaries ask: are the memories we experience actually ours? Our science-fictional imaginaries give us no reason to conclude either yes or no, leaving the answer no less certain than Deckard's status as human or Replicant at the conclusion of *Blade Runner.* Unlike the hopeful hybridity of Haraway's cyborg, the cyberpunk myth (epitomized by William Gibson's foundational novels) separates the "consensual hallucination" of cyberspace from "meatspace." This separation suggests that memory, once transferred to media outside of ourselves, is no longer ours. In fact, without memory, we appear no longer to be ourselves: memory scholars have described reminiscent modern forms of cultural amnesia and the loss of the art of memory in the rush of modern life. For example, Pierre Nora places history and memory in conflict, narrating the "conquest" of memory by an ever-accelerating historical record, a vast standing-reserve of documents piled skyscraper-high, burying the past in an act of archival "terrorism." These fears of "neuromantic" amnesia confront the "otherness" of prosthetic memory (Landsberg 2004; Csicsery-Ronay 1988). In a Cartesian schism of virtualized mind and subjugated meat, the cyberpunk myth depicts the self as "the victim . . . helpless and sad, against the powers of exteriorized mind" (Csicsery-Ronay, 277). The dual image of the cyborg—at once an enhancement of our embodied minds as well as the collapse of self and collective—confronts modern human uncertainties about the relationship of our bodies to the technologies that extend them.

Oblivion and the Archive

Discussions of digital memory often begin with the "memex" (memory-index), Vannevar Bush's 1943 vision of a permanent, electromechanical archive that would connect documents to each other by means of associative trails and annotations—much as hyperlinks do online today. Bush changed his position about whether memex documents should be permanent: while in his famous proposal all records are fixed, in his unpublished "Memex II" he stresses the need for "a readily alterable record" whose entries can be rewritten or deleted. This section outlines this continuum of records in light of digital memory media.

"Who controls the past controls the future," George Orwell famously remarked in *1984*, and "who controls the present controls the past." Alterability invites revisionism, whose purest incarnation is the infinitely alterable "memory hole" of *1984*'s dystopian society: an omnivorous control technology into which inconvenient documents are deposited and made to disappear or reappear at will, in the service of an official historical narrative. Specters of genocide and violence haunt us, imploring that we "never forget" even as the same imperative compels those who have suffered most to relive their traumatic pasts. Avishai Margalit calls for an "ethics of memory" that ensures descendants of genocidal trauma are not shackled to the ritual duty to commemorate without reflection (2002). His concerns echo Plato's condemnation of writing: that the historical text can only repeat its "one unvarying answer" to future questioners. Lacking the fluidity of conversation, a fixed record cannot evolve or learn; an alterable record, by contrast, offers hope for redemption, regeneration, and reconciliation in the possibility of radical change that "escapes" the past by rewriting it. Yet such historical revisions invariably reflect present-day values: both matters of public deliberation and knowledge production rehearse the ways individuals and collectives continually rewrite and resist rewriting our records of the past.

The question of how societies remember invites further questions: how and what do we archive? Which actors access and use data to construct cultural narratives? Jeffrey Rosen and Viktor Mayer-Schönberger argue for the "virtue" of forgetting as a rearguard defense against the threat of what they call permanent "comprehensive memory": "all citizens face the difficulty of escaping their past now that the Internet records everything and forgets nothing" (Mayer-Schönberger 2009; Rosen 2010). The persistent myth that "everything is recorded" online, however, fails to acknowledge its many actual "memory holes"—and, moreover, obscures the pressing questions of what is missing from the archive, and why. At the same time, they raise vital concerns about the effects of digital media on memory at the cultural and institutional levels: an expansion of surveillance and mass acquisition of data about individuals; a shift toward technological norms that, by default, record and retain such data; search engines that reveal individual data to the expert few or the trolling mobs; and the diminishment of the individual's ability to know or influence how those data are collected and used. According to Rosen and Mayer-Schönberger, digital archives that capture both imperfectly and ubiquitously our past preferences and actions compel legal protection of the faculty to forget—or, as in recent European contexts, the "right to be forgotten." Mayer-Schönberger, for example, proposes "expiration dates" on some types of archival data as a defensive maneuver against the perceived erosion of privacy and autonomy.

In fiction and real-life imagination alike, the search for perfect access to the past threatens to eradicate present privacy. In Asimov's allegory "The Dead Past," "comprehensive memory" and a "temporal panopticon" appear inseparable: a fictional "chronoscope," built for looking into the past, accidentally ends present autonomy to act without surveillance. Asimov writes: "The dead past is just another name for the living present. What if you focus the chronoscope in the past of one-hundredth of a second ago? Aren't you watching the present?" In Europe, *le droit à l'oubli*—the "right of oblivion"—signals our modern belief that past deeds and the present day can be kept separate through limitation of our memory media's chronoscopic creep. In contradistinction to the often hopeful view of representation in archival records ("going down in history") as an elixir of immortality, here the right to be forgotten ensures a fresh start—the chance to erase and escape traumatic elements of one's past.

Persistence, Ephemerality, and Power

In practice, neither imaginative extreme reflects technological fact: digital memory today is neither comprehensive nor permanent. "The World Wide Web still is not a library," concludes Wallace Koehler after conducting a longitudinal study of the "half-life" of online documents—and it is certainly not the universal informational repository imagined by the memex (2004). "Link rot" (hyperlinks whose destination pages are no longer available) introduces significant decay into the Internet's associative hypertext trails; "bit rot" and "data rot" similarly reveal that software degenerates as it accumulates naturally occurring errors on fragile hard disks and drives (digital memory has yet to stand the test of time); website providers can go out of business, causing thousands of pages to disappear overnight. The comparatively unindexed recesses of the "dark internet" and the "deep web" remind us that contemporary curation of archival data is anything but complete, comprehensive, or static.

Much of the Internet's content is still characterized by its practical ephemerality: the average thread on 4chan's popular /b/ message board, for example, spends a mere five seconds on the first page, and five minutes on the site in total, before its content vanishes (Bernstein et al. 2011). "Ephemeral technologies" like Snapchat, which delete information shortly after its receipt, implement forgetting, not remembering, as their digital default. While this inbuilt impermanence can be defeated by a "hack" as simple as a screenshot, the lack of a persistent, searchable digital data archive in such ephemeral platforms demonstrates how some new social norms favor records designed to be fleeting.

The temporal gaps and untraceable depths that characterize our memory media intensify rather than diminish the need for critical scrutiny of historical and archival practices. Our archives are curated, our histories continually constructed, and our traces of events incomplete. Individuals and institutions enjoy asymmetrical levels of control over their ability to access, analyze, and interpret information flows. These realities render David Brin's "transparent society," in which individuals and organizations have equal access to each other's data, unrealizable: in "postulat[ing] the end of privacy," contend Bossewitch and Sinnreich, "[the transparent society] fails to adequately account for the differential access to analytic processing power available to different individuals and organizations in making sense—and use—of this data" (2013).

Such power differentials shape archival lacunae. As Susan Brison argues, "As a society, we live with the unbearable by pressuring those who have been traumatized to forget and by rejecting the testimonies of those who are forced by fate to remember" (1996). Institutional power dynamics further inform who can seek and find reparation and forgiveness, who enjoys the hidden privileges of memory gaps, and who suffers the unbearable memories of others. As long as we do not live in a symmetrically transparent utopia, crucial concerns about selective cultural amnesia, surveillance, and power will press upon us: who controls the archives, the official histories that modulate collective memory? Who surveys the past, who is surveyed, and who can evade surveillance? How is information from the archive recalled, re-presented, recontextualized, and revised into new narratives about the past? Under what conditions are such narratives emancipatory or deceptive?

Conclusion: Memory as Surrogate or Symbiosis?

> Because we do not understand the brain very well we are constantly tempted to use the latest technology as a model for trying to understand it. In my childhood we were always assured that the brain was a telephone switchboard. ("What else could it be?") I

> was amused to see that Sherrington, the great British neuroscientist, thought that the brain worked like a telegraph system. Freud often compared the brain to hydraulic and electro-magnetic systems. Leibniz compared it to a mill, and I am told some of the ancient Greeks thought the brain functions like a catapult. At present, obviously, the metaphor is the digital computer.
>
> —John R. Searle, *Minds, Brains and Science*

> The way we conceive of natural symbol systems depends to a large degree on the computational metaphors we use to understand them, and machine learning suggests an understanding of symbolic thought that is very different to traditional views. . . . Our analysis of [predictive, probabilistic symbolic communication] arose out of the idea that the mind can be modeled as a kind of learning machine.
>
> —Michael Ramscar, "Computing Machinery and Understanding"

No memory technology is immune to the influence of its own history (Kalnikaitė and Whittaker 2007); even history itself arises out of the development of the medium of writing. All memory is both mediated and historically contingent. Nonetheless, we can conclude this necessarily speculative framework for thinking about memory in the modern age with a few perennial and pressing questions: Whose are our memories? Do memory media embody or displace our imagined "selves"? Does embodied memory lead to cultural symbiosis, as Plato suggested, while prosthetic memory leads to a technological surrogacy in which external actors and archives deputize the incomplete reinvention of our identities—and how do we construct and maintain distinctions between the two?

As Searle illustrates, our metaphors for that seat of memory, the mind, have historically modeled cognition as a pneumatic system, a clockwork automaton, a helmsman steering a ship, an enchanted mechanical loom; today, we breezily compare the mind to a search engine, algorithmically retrieving stored data from a disorganized network, "learning" from each new search of its archive. As we employ this new media metaphor of memory and use it to reimagine both our individual and our collective identities, we also derive meaning from the mind-metaphors of past eras, and the politics and poetics of their memories preserved. Contemporary "cloud computing" extends cyberpunk notions of disembodied mind into the present day, while proliferating augmented- and virtual-reality technologies reinvigorate hopes and fears about the potential of the cyborg to emancipate or dehumanize. While these old concerns accompany every new paradigm, they yet remain relevant: digitally reconstructed memory and forgetting continue to exist in a state of paradox and plurality, inviting continued conversation about our archives, our histories, and ourselves.

New mnemonic technologies revitalize timeless questions about the contradictory nature of memory—constantly reconstructing the past while prospecting potential futures, in acts as simple as reading old letters from a friend or writing a shopping list—and resurrect familiar specters as well. But as individuals and corporations alike increasingly seek out professional reputation management services to influence their archival afterimages (at least, those who can afford to do so), and the European Court of Justice navigates the tension between privacy and free expression implicated in a (limited) "right to be forgotten" from the index of search engines, these questions and anxieties gain urgency and force. By tracing prevalent themes of information control, surveillance, and power against the background of prosthetic memory, this keyword comes to represent more than either synapses or hard disks—far from signaling either the end of memory or the end of forgetting, our shifting metaphors for memory and mind represent the complex and multivalent influence of the present upon the future and past.

See in this volume: archive, cloud, community, event, forum, meme, mirror, prototype, surrogate

See in Williams: alienation, bureaucracy, civilization, collective, consensus, conventional, dialectic, experience, history, interest, media, modern, representative, society, tradition

References

Bernstein, M. S., et al. 2011. "4chan and/b: An Analysis of Anonymity and Ephemerality in a Large Online Community." ICWSM.
Biddick, K. 1993. "Humanist History and the Haunting of Virtual Worlds: Problems of Memory and Rememoration." *Genders*, no. 18: 47–66.
Biocca, F., and M. R. Levy. 1995. *Communication in the Age of Virtual Reality*. Hillsdale, NJ: Lawrence Erlbaum Associates.
Bossewitch, J., and A. Sinnreich. 2013. "The End of Forgetting: Strategic Agency beyond the Panopticon." *New Media & Society* 15(2): 224–42.
Bowker, G. C. 2005. *Memory Practices in the Sciences*. Cambridge, MA: MIT Press.
Brison, S. J. 1996. "Outliving Oneself: Trauma, Memory and Personal Identity." In *Feminists Rethink the Self*, ed. Diana T. Meyers. Boulder, CO: Westview Press.
Bugeja, M., and D. V. Dimitrova. 2006. "The Half-Life Phenomenon." *The Serials Librarian* 49(3): 115–23.
Caruth, C. 1996. *Unclaimed Experience: Trauma, Narrative, and History*. Baltimore: Johns Hopkins University Press.
Cavallaro, D. 2000. *Cyberpunk and Cyberculture: Science Fiction and the Work of William Gibson*. New Brunswick, NJ: Athlone Press.
Chun, W.H.K. 2008. "The Enduring Ephemeral, or the Future Is a Memory." *Critical Inquiry* 35(1): 148–71.
Connerton, P. 1989. *How Societies Remember*. Cambridge: Cambridge University Press.
———. 2009. *How Modernity Forgets*. Cambridge: Cambridge Unoversity Press.
Csicsery-Ronay, I. 1988. "Cyberpunk and Neuromanticism." *Mississippi Review* 16(2/3): 266–78.
Daugman, J. G. 2001. "Brain Metaphor and Brain Theory." In *Philosophy and the Neurosciences: A Reader*, ed. William P. Bechtel, Pete Mandik, Jennifer Mundale, and Robert S. Stufflebeam. Malden, MA: Blackwell.
Derrida, J. 1995. "Archive Fever: A Freudian Impression." *Diacritics* 25(2): 9–63.
Ernst, W., and J. Parikka. 2012. *Digital Memory and the Archive*. Minneapolis: University of Minnesota Press.
Featherstone, M., and R. Burrows. 1996. *Cyberspace/Cyberbodies/Cyberpunk: Cultures of Technological Embodiment*. London: Sage Publications.
Foucault, M. 1977. *Language, Counter-memory, Practice: Selected Essays and Interviews*. Ithaca, NY: Cornell University Press.
Garde-Hansen, J., et al. 2009. *Save as . . . Digital Memories*. New York: Palgrave Macmillan.
Haskins, E. 2007. "Between Archive and Participation: Public Memory in a Digital Age." *Rhetoric Society Quarterly* 37(4): 401–22.
Kalnikaitė, V., and Steve Whittaker. 2007. "Software or Wetware? Discovering When and Why People Use Digital Prosthetic Memory." In *Proceedings of the ACM CHI 2007 Conference on Human Factors in Computing Systems*, 71–80. San Jose, CA: ACM.
Keightley, E., and M. Pickering. 2012. *The Mnemonic Imagination: Remembering as Creative Practice*. New York: Palgrave Macmillan.
Kennedy, V. 1999. "The Computational Metaphor of Mind: More Bugs in the Program." *Metaphor and Symbol* 14(4): 281–92.

Koehler, W. 2004. "A Longitudinal Study of Web Pages Continued: A Report after Six Years." *Information Research* 9(2): 1–9.
Landsberg, A. 2004. *Prosthetic Memory: The Transformation of American Remembrance in the Age of Mass Culture*. New York: Columbia University Press.
Luckhurst, R. 2008. *The Trauma Question*. London: Routledge.
Margalit, Avishai. 2002. *The Ethics of Memory*. Cambridge, MA: Harvard University Press.
Markwell, J., and D. W. Brooks. 2003. "'Link Rot' Limits the Usefulness of Web-Based Educational Materials in Biochemistry and Molecular Biology." *Biochemistry and Molecular Biology Education* 31(1): 69–72.
Mayer-Schönberger, Viktor. 2009. *Delete: The Virtue of Forgetting in a Digital Age*. Princeton, NJ: Princeton University Press.
McGlone, M. S. 2007. "What Is the Explanatory Value of a Conceptual Metaphor?" *Language & Communication* 27(2): 109–26.
Muri, A. 2003. "Of Shit and the Soul: Tropes of Cybernetic Disembodiment in Contemporary Culture." *Body & Society* 9(3): 73–92.
Nichols, B. 1988. "The Work of Culture in the Age of Cybernetic Systems." *Screen* 29(1): 22–46.
Parker, A. 2007. "Link Rot: How the Inaccessibility of Electronic Citations Affects the Quality of New Zealand Scholarly Literature." *New Zealand Library & Information Management Journal* 50(2): 172–92.
Pruchnic, J., and K. Lacey. 2011. "The Future of Forgetting: Rhetoric, Memory, Affect." *Rhetoric Society Quarterly* 41(5): 472–94.
Ramscar, M. 2010. "Computing Machinery and Understanding." *Cognitive Science* 34(6): 966–71.
Rosen, J. 2010. "The Web Means the End of Forgetting." *New York Times*, July 21.
Ruiz de Mendoza Ibáñez, F. J., and L. Pérez Hernández. 2011. "The Contemporary Theory of Metaphor: Myths, Developments and Challenges." *Metaphor and Symbol* 26(3): 161–85.
Saerle, J. 1984. *Minds, Brains, and Science*. Cambridge, MA: Harvard University Press.
Sellen, A. J., and S. Whittaker. 2010. "Beyond Total Capture: A Constructive Critique of Lifelogging." *Communications of the ACM* 53(5): 70–77.
Steiner, L., and B. Zelizer. 1995. "Competing Memories: Reading the Past against the Grain: The Shape of Memory Studies." *Critical Studies in Mass Communication* 12(2): 213–39.
Sternberg, R. J. 1990. *Metaphors of Mind: Conceptions of the Nature of Intelligence*. Cambridge: Cambridge University Press.
Van Dijck, J. 2007. *Mediated Memories in the Digital Age*. Stanford, CA: Stanford University Press.
———. (2013). "'You have one identity': Performing the Self on Facebook and LinkedIn." *Media, Culture & Society* 35(2): 199–215.
Van House, N. and E. F. Churchill. 2008. "Technologies of Memory: Key Issues and Critical Perspectives." *Memory Studies* 1(3): 295–310.
Viégas, F. B., et al. 2004. "Digital Artifacts for Remembering and Storytelling: Posthistory and Social Network Fragments." In *Proceedings of the 37th Hawaii International Conference on System Sciences*. IEEE.
Wiener, N. 1988. *The Human Use of Human Beings: Cybernetics and Society*. Boston: Da Capo Press.

20

Mirror

Adam Fish

The mirror is one of the most copied metaphors in Western thought. In ancient Greek mythology, Narcissus dies transfixed on his reflection in a spring. According to early sociology, we are a "looking-glass self." Our identities are formed when we mirror how we think others see us (Cooley 1902). In *Philosophy and the Mirror of Nature*, Richard Rorty (1979) shatters the Enlightenment ideal that through scientific inquiry the mind could mirror nature, harboring replicas in mental formulas. Thus, from antiquity onward, the mirror metaphor has been used to describe everything from vanity, to subject formation, to consensual reality. Today, information companies and information activists alike call data duplication *mirroring* but often fail to acknowledge how the symbolism of this term may impact its use. Mirrors are more complex and faulty entities than simple facsimiles. With duplications come decreasing fidelity and increasing glitch. As social processes, *mirrors* echo the intricacies and limitations of data practice. I endeavor to explain how information activists and information firms make mirrors in order to exploit computer networks and remain visible.

Mirrors—derived from the Latin *mirare* for "to look at"—are metaphors for what they reflect. In *Through the Looking-Glass*, Lewis Carroll (1871) has Alice journey through a mirror and into a parallel and parable-rich universe of reversals. In Oscar Wilde's *The Picture of Dorian Gray* (1891), the mirroring portrait ages but the protagonist does not. Hillel Schwartz (1996) traces this history and our obsession with twins, replicas, duplicates, decoys, counterfeits, portraits, mannequins, clones, replays, photocopies, and forgeries. The mirror metaphor continues into the digital age. The United Kingdom's Channel Four television series *Black Mirror* is a drama that comments on a dystopic future of increasing connectivity. Charlie

Booker's program sees our mobile and laptop screens as black mirrors into which we stare Narcissus-like and which reflect back our self-destructive ways. Peter Sunde, cofounder of file-sharing company The Pirate Bay, spent time in prison for his copying acts and believes that copying is genetically coded, saying: "People learn by copying others. All the knowledge we have today, and all success is based on this simple fact—we are copies" (Ernesto 2014). As a locus for the confluence of metaphysics and materiality, mirrors are a way to see how the practical and metaphoric are coconstituted in database worlds.

For Jacques Lacan the mirror stage is the moment between six and eighteen months when a child apperceives or objectifies her subjectivity. This turning inside out is an externalization of interiority, or the freezing of the modern subject. Referencing the technology of his time, Lacan invokes a jammed cinema projector that is suspended on a single frame that then becomes the ego (Žižek 1997). Identification begins and, for Lacan, alienation and narcissism soon follow: the mirror is no longer a stage but an imaginary and fraudulent state that permanently masks the absence of the symbolic and the unattainability of the real. Scholars following late Lacan, by contrast, extend the metaphor to describe the mirror as the site of the formation of the subject, where the virtual is an ideal made real that emerges from "games of mirroring" (Deleuze 1972, 172). Likewise data copies or mirrors are virtual ideals of perfect duplication made imperfectly real. They are frozen information externalizations, duplications that strive for unattainable states of exacting verisimilitude. Referencing the technology of our time, we can think of a database mirror as a replication of a frozen operating system—the Apple "spinning pinwheel" or, less formally, "spinning beach ball of death"—that locks a user's screen into an imperfect frieze.

This essay seeks both to discuss data mirrors as a metaphor as well as to show how mirroring serves as a practice of data activism and cloud computing. Below I describe how computing mirroring keeps a copy of some or all of a particular content at another remote site, typically in order to protect and improve its accessibility. Mirroring multiplies data sources. For activists, mirroring is a method to preserve and protect visibility through duplicating and distributing their resources across communication networks. Mirror multiplicities also allow cloud companies to capture and sell personal information. Geographically dispersed and intensely complex, mirrors are no simple replication of origins: rather they are a form of praxis or a way of being and becoming in the networked world. Data mirroring reveals in our digital reflections a hall of mirrors between the practical and the metaphoric, the actual and the virtual, the hyperreal and the ideal.

Both the replication and visibility elements of mirroring are political. Reflecting, repeating, amplifying, translating, replicating, and copying are core modes for understanding the control of modern information. These practices are often but not necessarily visual. In computational culture, the seen and the unseen are interlinked in ways not easily perceived. Mirrored databases, XML spreadsheets, copied JPEGs, and torrented videos each have visual components allowing front-end users to graphically interface with back-end code. In this way, screened, front-end interfaces translate computer applications for human readability. The front end is what is visible, seen, public, and, as a semantic object, most easily subjected to political deliberation and economic capture. The back end, where a hidden battle for control and capture of information is waged, is invisible to all except expert engineers and hackers (see **hacker**).

Mirroring data sets from private and invisible sites to public and visible ones often renders such battles visible. Mirroring punctures with data leaks the veil hiding the back end, so that the links between the visible front end and invisible back end too are made visible. The machinery is exposed and the black box of hardware opened. In this way, replication of remote data sets becomes a question of visibility. One struggle is about control over the back end and privacy; another is focused on who has the capacity to make the invisible visible in public. While I emphasize the visual front end of mirrored data sets, it is the mirroring or duplication of the back-end data and metadata that drives understanding of what is possible with the digital. Mirroring is a unique and contemporary digital manifestation of that always politicized act of information replication.

Mirrors as Multiples

Mirrors are multiples. Mirroring serves several purposes. Cloud computing relies on mirroring or replication of databases for global access and security. Microsoft, which provides a number of cloud computing services, defines "database mirroring" as the maintenance of "two copies of a single database that must reside on different server instances." The basic copy-and-paste function of networked digital computing makes possible, according to these same computing companies, the nonrivalrous multiplication of data. Not only Fortune 500 information companies marshal mirroring techniques to preserve and protect their data integrity. Data and transparency activists with WikiLeaks also actively "mirror" its content. They and their supporters mirror content in jurisdictions outside America in the face of the legal shutdown of private servers housing their incendiary leaks. Today, servers in at least eleven European nations offer the WikiLeaks mirror (http://wikileaks.info/). The largest peer-to-peer file-sharing service in the world, The Pirate Bay, mirrors its links on servers in national jurisdictions where its practices have yet to be deemed illegal (eighteen countries presently block root access to The Pirate Bay). Mirroring, thus, is a replication practice for both hegemonic and counterhegemonic actors. Despite this political symmetry, it would be misleading to claim that mirroring produces exact replicas.

To offer robust, secure, and nondelayed access to content, it is necessary to store multiples. Yet Microsoft offers a naive realist notion that mirrors are precise copies, merely displaced within or across databases. A slightly more complex social constructivist perspective sees mirrors as symbolic representations. In constructivism, mirrors would be conceived not as duplicates but rather as iconic yet accurate depictions. Physicist Karen Barad (2003) challenges both "naive realist" as well as constructivist interpretations of mirrors, offering a third construal. Echoing Rorty, she says, "The representationalist belief in the power of words to mirror preexisting phenomena is the metaphysical substrate that supports social constructivist, as well as traditional realist, beliefs" (Barad 2003, 802). Mirrors produce neither realist copies nor constructed depictions. Rather mirrors are data multiplications that make political contests visible.

In other words, data mirroring does not represent so much as it reveals the complexities of those who mirror their content. For example, in cloud computing, content is retrieved and recomposed from geographically remote databases connected by complex networks. Instead of representing these networks as single entities, we should visualize them as similar to other complex networks, such as disease, criminal, and biological processes in nature. Each multiple, whether a mirrored file or a wild virus, exists in its numerous coded transactions (Ruppert 2013; MacKenzie and McNally 2013). In each case, the multiple is no fragmented or contradictory singularity. It is a fluid "field of multiple conjoined actions that cumulatively enact new entities" (Ruppert 2013). The "performative excesses" of multiples "undo or unmake identities as much as they make them" (Mackenzie and McNally 2013). Structured by diverse databases and unmoored from single origins, mirrors are multiples that serve hegemonic and counterhegemonic actors alike.

Mirroring as Activist Visibility

Mirrors transform seeing and what is seen. The legitimacy of WikiLeaks, The Pirate Bay, and Anonymous, among other counterhegemonic forces, rests on their ability to remain seen through replication. This is of course nothing new. Through physical vanity mirrors, people of medieval Europe "came to reflect on, know and judge themselves and others through becoming aware of how they appeared" (Coleman 2013, 5; Melchior-Bonnet 2001). Using lenses and mirrors to transform his studio into a camera obscura, seventeenth-century Dutch painter Johannes Vermeer depicted not the depth of field and the textures seen by the unmediated human eye but the world as framed by a camera obscura (Steadman 2002). Herein lies another regime of technological-assisted seeing and copying. Historically, writing and printing systems prioritized and privileged the ocular (or what the eye could see), allocating power to those who could read, write, print, and evaluate based on text (Ong 1977; McLuhan 1964). "Scopic regimes," such as Western science and law, control power by making certain things visible and legible, and others not (Jay 1992). Likewise, visual technologies organize and assemble the real, the natural, and the moral for Western technoscientific systems (Haraway 1997). By "seeing like a state," nations objectify and thereby control colonial bodies (Scott 1999). This will to visibility is also profoundly gendered in cinema that has historically served the male gaze (Mulvey 1975). Visibility "lies at the intersection of the two domains of aesthetics (relations of *perception*) and politics (relations of *power*)" (Brighenti 2007, 324). So too are digital mirrors replications of files in order to manipulate both their legibility and their legitimacy. Even the term *replicate* means etymologically "to fold back": and to fold back, or to ply (re-pli-cate) something, suggests such literal manipulation (see **digital**).

Leaking classified information is obviously a political and, in some countries, treasonous act. That copying could be as inherently political is less obvious, however. Lisa Gitelman, for example, emphasizes not the leaking but the photocopying of the Pentagon Papers by whistle-blower Daniel Ellsberg as the duplication strategy to make visible the invisible (Gitelman 2011). Contrasting the slow analog act of duplicating thousands of sheets of paper to the instantaneous work of WikiLeaks in which the "entire site was also 'mirrored' in several places around the world" (2011, 122) Gitelman sees a return to an older activism of making visible through duplication—a "*glasnost redux*" (Gitelman 2011, 122).

Or consider how Anonymous—made famous by hacks, leaks, and performative politics—secures visibility and subtle marketing for their political videos by mirroring them across YouTube. Gaines (1994) calls the process by which political videos hail viewers to copy revolutionary subjects "political mimesis." Again, following Barad, while mirrors represent politicized bodies, they are more than mere representations. Here, mirrors do not reveal sources but rather reveal conflict and contestation. For example, in response to the Church of Scientology's attempt to force YouTube to take down earlier Anonymous videos critical of Scientology, Anonymous decided to mirror its videos on YouTube. Instead of representing or responding to those who resist their criticism, Anonymous appeals to video mirroring as a way to make visible the conflict itself. The mirrors of Anonymous proliferate and preserve videos so that they might motivate others to mimic the radical personae represented in the videos (Fish 2015). The mirrored videos do not mark just the videos. Each activist video mirror reveals a once-hidden conflict by simulating the conflict it cites. Hacks, leaks, and video mirrors are forms of visual counterpower. The power to see and not be seen—from the eye training of literacy, to the male gaze in cinema, to cultures of self-presentation and reality television, to visibility-optimization industries of fashion and advertising, to video mirroring—constitutes regimes of power and counterpower in networked society.

When activist groups such as WikiLeaks, Anonymous, and The Pirate Bay mirror, their radical politics cannot help but "misuse" capitalist information infrastructure (Soderberg 2010). Despite their reliance on for-profit social media platforms (Dean 2010), grassroots mirroring still raises voices that resist censure in the circuits of technocapitalism (Couldry 2010). Mirroring is one among many promising but nonetheless uneven forms of technological resistance available to support and resist for-profit capture of information.

Capturing the Mirror

The short story of human history may be told as one based on the incremental accumulation of information (Gleick 2011). Human evolution is marked by a slow collective increase in the size and complexity of the neocortex, language, group dynamics, and information exploitation (Dunbar 1993). Likewise, the data-carrying capacity of media materiality—rock, wood, pulp fiber, electrical pulses, and now digital circuits—to store, transmit, and process symbolic systems has dovetailed with the increasing complexity of the brain, language, and society (Ong 1977; McLuhan 1964). Mirroring is just one among many manifestations of the prehistoric practice of data communication, control and manipulation—and, as ever before, current institutional risks and the political economy of corporately owned databases shape and structure the states of virtual data. The "double-movement of liberation and capture" (Deleuze 1989, 68) lets activists appear visible in public at the same time they as let data corporations capture social capital. Simultaneously, distributed mirroring allows activist data to escape capture on sites corporations do not own.

The business proposition of cloud companies is that mirroring is an affordable and socially responsible way of securing retrievable data. We congratulate ourselves when we back our data up, post autobiographical and personal artifacts, and work on the go by placing our documents in the cloud. The same proposition is compromised, however, by the fact that all this plays into surveillance with unseen consequences and costs for our body politic. Bound up in the back-and-forth of hegemonic and counterhegemonic power struggles, mirroring is no innocent activity: it captures for some and liberates for others the very data it displaces, diffracts, and makes autonomous. It serves activist visibility as well as a trap of the same.

Conclusion

Mirrors make and save copies in different places. But, as we all know, mirrors make no exact copies and identify not with their reflections. Mirrors are not products but rather idiosyncratic processes for creating complex multiples autonomous from their origins. Mirroring does not promise realistic representations. Rather, it offers a way of being, acting, and moving in the world. Mirrors map and reveal both activist and corporate forms of conflict and contestation. For activists, mirroring reveals a will to remain visible in a world of censorship, surveillance, and information infrastructural control. For cloud companies, mirroring marks conflicts over the capture and capitalization of data. Mirrors—understood as sites for making and liberating multiples—synthesize key elements of modern informational political economy and praxis. We have rarely been good at facing our doubles: Narcissus dies of starvation by the edge of a pool, Dorian Gray lacerates his mirror painting and stabs himself in the heart, and so too is modern integrity put in peril by the proliferation of the copies of our many selves. That said, what happens behind the mirror—in the invisible back end that manages metadata and structure—may be more contentious than what happens in front of it.

Duplicated files are often political (e.g., counterfeit Roman coins, Lutherian theses, East German facsimiles, Xeroxes of the Pentagon Papers), but data mirroring suggests a new contentious hidden infrastructure for duplicating and distributing data and their identifying metadata. The metadata intensifies the politics of mirroring, since every act can be seen by some and hidden from others. Legal struggles have accelerated over the battle to control and reform peer-to-peer networks and copyright regimes. Overzealous prosecution of open-culture activists drove the tragic suicide of Aaron Swartz, who copied academic journal articles to mirrored databases. Copying and propagating classified documents, Julian Assange and Edward Snowden both speak to the profound visibility of recent mirroring activism. The problems and potentials of mirroring are unlikely to disappear anytime soon. Mirroring belongs to an ancient tradition of replication. Once reserved for scribes and technicians, copying-and-pasting has become perhaps the most powerful practice in everyday computing. Mirroring magnifies the significance of transparency, openness, and visibility—*glasnost redux*, indeed.

See in this volume: analog, cloud, culture, digital, flow, hacker, internet, memory, surrogate

See in Williams: aesthetic, behavior, bureaucracy, capitalism, charity, collective, common, consumer, exploitation, idealism, labour, management, media, organic, popular, society, taste, technology, work

References

Barad, Karen. 2003. "Posthumanist Performativity: Toward an Understanding of How Matter Comes to Matter." *Signs: Journal of Women in Culture and Society* 28(3): 801–31.

Brighenti, A. M. 2007. "Visibility: A Category for the Social Sciences." *Current Sociology* 55(3): 323–42.

Coleman, Rebecca. 2013. "Sociology and the Virtual: Interactive Mirrors, Representational Thinking and Intensive Power." *Sociological Review* 61(1): 1–20.

Cooley, Charles H. 1902. *Human Nature and the Social Order*. New York: Scribner's.

Couldry, Nick. 2010. *Why Voice Matters*. London: Sage Press.

Dean, Jodi. 2010. *Blog Theory*. Cambridge: Polity Press.

Deleuze, Gilles. 1989. *Cinema 2: The Time-Image*. Translated by Hugh Tomlinson and Robert Galeta. Minneapolis: University of Minnesota Press.

Dunbar, R.I.M. 1993. "Coevolution of Neocortical Size, Group Size and Language in Humans." *Behavioral and Brain Sciences* 16(4): 681–735.
Ernesto. 2014. "Pirate Bay Founder Gets Ready to Run for European Parliament." *Torrentfreak*. https://torrentfreak.com/pirate-bay-founder-gets-ready-to-run-for-european-parliament-140321/.
Fish, Adam. 2015. "Mirroring the Videos of Anonymous: Cloud Activism, Living Networks, and Political Mimesis." *Fibreculture Journal* 25.
Gaines, Jane M. 1994. "Political Mimesis." In *Collecting Visible Evidence*, edited by Jane M. Gaines and Michael Renov. Minneapolis: University of Minnesota Press.
Gitelman, Lisa. 2011. "Daniel Ellsberg and the Lost Idea of the Photocopy." In *History of Participatory Media, Politics and Publics, 1750–2000*, edited by Andrew Ekstrom, Solveig Julich, Frans Lundgren, and Per Wisselgram, 112–24. New York: Routledge.
Gleick, James. 2011. *The Information: A History, a Theory, a Flood*. New York: Pantheon Books.
Haraway, Donna. 1997. *Modest_Witness@Second_Millennium.FemaleMan© Meets_OncoMouse™: Feminism and Technoscience*. New York: Routledge.
Jay, Martin. 1992. "Scopic Regimes of Modernity." In *Vision and Visuality*, ed. Hal Foster, 3–23. Seattle, WA: Bay Press.
Mackenzie, Adrian, and Ruth McNally. 2013. "Methods of the Multiple: How Large-Scale Scientific Data-Mining Pursues Identity and Differences." *Theory, Culture & Society* 30(4): 72–91.
McLuhan, Marshall. 1964. *Understanding Media*. New York: McGraw-Hill.
Melchior-Bonnet, Sabine. 2001. *The Mirror: A History*. London: Routledge.
Mulvey , Laura. 1975. "Visual Pleasure and Narrative Cinema." *Screen* 16(3): 6–18.
Ong, W. J. 1977. *Interfaces of the Word: Studies in the Evolution of Consciousness and Culture*. Ithaca, NY: Cornell University Press.
Rorty, Richard. 1979. *Philosophy and the Mirror of Nature*. Princeton, NJ: Princeton University Press.
Ruppert, Evelyn. 2013. "Not Just Another Database: The Transactions That Enact Young Offenders." *Computational Culture*, November 16, 1–13.
Schwartz, Hillel. 1996. *The Culture of the Copy: Striking Likenesses, Unreasonable Facsimiles*. New York: Zone Books.
Scott, James. 1999. *Seeing Like a State: How Certain Schemes to Improve the Human Condition Have Failed*. New Haven, CT: Yale University Press.
Soderberg, Johan. 2010. "Misuser Inventions and the Invention of the Misuser: Hackers, Crackers, Filesharers." *Science as Culture* 19(2):51–79.
Steadman, Phillip. 2002. *Vermeer's Camera: Uncovering the Truth behind the Masterpieces*. Oxford: Oxford University Press.
Žižek, Slavoj. 1997. *The Plague of Fantasies*. London: Verso.

21

Participation

Christopher Kelty

Participation is like a monument one passes every day—so routine, so common it's hard to remember just why it is there, or what it memorializes. When we read a Wikipedia page, answer an Android phone, or log into Instagram, it is easy to, so to speak, *walk past* this monument. The experience of such mediated collectives, and the fact of their material existence, owes much to the keyword *participation*—to the many real hands and minds and hearts necessary to produce that image, that encyclopedia entry, that emotional experience of belonging. Even in a volume of digital keywords such as this, one can risk missing its centrality unless someone points it out, a bit like an annoying tour guide bent on civic pride, detouring away from charismatic concepts like *democracy*, *hacking*, or *sharing* as the crowded, unmissable, tout-ridden destinations at the heart of the city. But there the concept hides in plain sight, with its peculiar power: "One cannot say we did not notice [it]; one would have to say [it] 'de-notices' us" (Musil 1987).

Participation has lately been singled out as one of the key features of contemporary digital culture (Carpentier 2011; Delwiche and Henderson 2013; Deuze 2006; Jenkins et al. 2007) whether to signal the promise of something positive (a more egalitarian, just, democratic participation) or to signal a failure (a more exploitative, extractive, involuntary form of participation). From 1994 to 2015, the term found itself applied, with an intensely renewed vigor, to the Internet, new media, mobile technologies, and social media: from free software, to Web 2.0 to social media to the Arab Spring, *participation* was a keyword often invoked in combination with *Internet*, *democracy*, and *freedom*. In 2010, for instance, Malcolm Gladwell issued a *New Yorker* salvo suggesting that digital social media participation (the "Arab Spring" or the Iran "Twitter revolution") is only an attenuated and low-stakes version of "real" participation (such as the civil rights sit-ins and marches of the 1960s) (Gladwell 2010). This kind of move is a clear indication of the *aspirational* aspect of the word—it is almost always critiqued in the name of a better, more authentic participation to come. It is hard for anyone other than dictators, autocrats, and Bartleby to be against participation (Casemajor et al. 2015).

In terms of the last two decades of enthusiasm, utopianism, and criticism in and around things digital, the practices surrounding participation have played a central role. The rise of free and open-source software and user-led innovation explicitly focused on new forms of autonomous task choice and "Bazaar"-style (bottom-up) instead of Cathedral-style (top-down) organization of engineering and design (Chesbrough 2003; Hippel 2005; Raymond 1999). The label *peer production* singles out the new centrality of emergent collaborative forms of production and distribution of information goods over against hierarchies and markets of the past; Wikipedia is its emblem, but many other examples were paraded about as new ways of organizing production and consumption (Benkler 2006; Bruns 2008; Tkacz 2015). Crowdsourcing and crowdfunding evoked a new mode of aggregation or emergence of collective wisdom of participants used to solve problems, direct investments, or guide organizations and decisions. Fan fiction and "participatory culture" have exploded in popularity, moving from obscure, illegitimate, or subcultural spaces to mainstream, profitable media, and to new forms of civic involvement (Fish 2013; Jenkins 1992). And last but not least, the rise of "Big Data" alongside WikiLeaks and Snowden's revelations of spying, secrecy, and surveillance has revealed the role of *involuntary* participation, where the very idea of *not* participating in the digital has become an impossible or at least uncertain alternative.

Participation, n.

Participation's vagueness no doubt relates to its generality. As a word it is quotidian and unfussy:

> **1a** The action or fact of having or forming part *of* something; the sharing *of* something. In early use: the fact of sharing or

> possessing the nature, quality, or substance *of* a person or thing. (*Oxford English Dictionary* online, s.v. participation)

Its earliest appearance in English is in the nominative form—but it is much more common as a verb: "to take part; to have a part or share with a person, in (formerly also of) a thing; to share" (ibid.). It clearly overlaps with the language of sharing, although more often as an action than as a thing. We share food, but we participate in dinner, or eating. Indeed, participation has many partial cognates: engagement, collaboration, cooperation, involvement, democratization, or sharing. But a key feature of the concept is that participation comes with an effect:

> 3 The process or fact of sharing in an action, sentiment, etc.; (now esp.) active involvement in a matter or event, esp. one in which the outcome directly affects those taking part. (ibid.).

The benefit or effect of participation is most often understood to accrue to the participant herself. For example, one receives an "educative dividend" of learning how government works by participating in it. But participation can also benefit the entity enabling it—as when a corporation makes "worker participation" a key platform in measuring productivity or efficiency; or when scientists can claim advances and discoveries that depend on the contributions of countless observers, as in astronomy or ornithology (Wiggins and Crowston 2011; Vetter 2011). Almost all contemporary digital production and communication involving participation invokes this "dyadic" meaning—designating a smaller formal entity that both provides and receives benefit from a much larger, inchoate public or market of participants (Fish et al. 2011).

Heteronym

A peculiar feature of the term *participation* is that the same word seems to have multiple, nonoverlapping uses. These "heteronyms" have their own distinct discursive and epistemic circuits covering a remarkably wide range of inquiries, experiments, and domains of practice that are both practical problems to be solved and theoretical conundrums for scholars. The adjectival form, *participatory*, is an evocative and familiar case in point. It is the offspring of a particular time and place: the coining of "participatory democracy" in the 1962 Port Huron Declaration of the Students for a Democratic Society in Michigan, written by Tom Hayden (Hayden 1962). Participatory democracy, as a slogan, ascended rapidly, and the adjective has since been used to modify just about every form of human endeavor. Subsequently, one can find participatory art, budgeting, culture, design, economics, learning, management, medicine, planning, research, and urbanism, just to name the most obvious. *Participatory* shares some of the vagueness and rhetorical sloppiness of *excellence* in contemporary culture—signaling a nostalgia and a normative desire without risking any specific promises.

An early exemplar of participation as a practical problem is the creation of cooperatives starting in the 1830s and 1840s, and especially the famed "Rochdale Pioneers" and their principles of 1844, which include rules about membership, duty, sharing of resources through dividends, and the role of education. Such co-op style organizations have been a persistent, if periodic, feature of capitalist economies ever since (Cole 1944).

Participation gained a new momentum under the label *Industrial Democracy* in nineteenth-century Britain and America. Fabian socialists in Britain and early institutional economists in America (such as Richard Ely) understood participation as the extension of representative democracy from the state to the workplace, and possibly to all domains of life (Lichtenstein 1993; Derber 1970). This approach actualizes a specific idea of democracy (Lincoln's "of the people, by the people and for the people"), wherein the everyday "participatoriness" of society is related to the overall inclusiveness and democratic nature of a particular state and sovereign people (as opposed to the "elite" theories of representative democracy discussed below (see also **democracy**). The piecemeal replacement of industrial democracy by unions and collective bargaining after World War I also exemplifies one of the tensions in participation—that it often denies the existence of antagonism and functions instead as a "containment strategy" rather than an acknowledgment of class struggle.

In the 1950s and 1960s, groups such as the Situationists or Allen Kaprow and his "happenings" inaugurated a tradition of participation in conceptual and avant-garde art (Frieling 2008; Turner 2013) that has received its most enthusiastic expression and critique since the 1990s, when "relational aesthetics" or "socially engaged" art dominated the activities of artists like Felix Gonzales-Torres, Tino Seghal, Helio Oiticia, Rirkrit Tiravanija, or Jeremy Deller (Bourriaud 2002; Bishop 2012). Here participation has evoked a particular question about the autocracy of the artist and the question of aesthetics and its compatibility with social change or activism.

Participation appears forcefully and even doctrinally (e.g., "maximum feasible participation" was mandated in the War on Poverty legislation) in the Great Society of 1960s America. Here, participation was extended beyond issues of legislation and popular will to the very administration of government. It moved beyond the hands of elite experts and into the very citizens who would build, for instance, "Model Cities" and transform communities (Arnstein 1969; Haar 1975). Beginning with these experiments, "direct citizen involvement" has expanded and bureaucratized administrative activities, perhaps most visibly in cases of environmental regulation (Beierle and Cayford 2002) and more recently in attempts to involve the public in science and technology policy (Jasanoff 2003; Fisher, Mahajan, and Mitcham 2006; Wynne 2007; Irwin and Wynne 1996; Rowe 2005).

In the wake of the "participatory democracy" enthusiasms of the 1960s, worker participation saw renewed attention and experimentation throughout the 1970s and 1980s, ranging from the novel experiments of "Scandinavian Participatory Design" (Asaro 2000), initially concerned with control over the introduction of technology in factories (and more recently the inclusion of client perspectives in the design disciplines), to the proliferation of management theories about "teams," "total quality management," or "high-involvement" workplaces designed to benefit both efficiency and worker "satisfaction" (Miller and Rose 1988; Boltanski and Chiapello 2005; Lezaun 2011).

As a last example, participation has been most thoroughly incorporated—and critiqued—in the domain of international development since the 1980s (Cooke and Kothari 2001). The United

Nations, the World Bank, and other nongovernmental organizations have advanced participation practices such as participatory rural appraisal (PAR), promulgated evangelically by Robert Chambers (Chambers 2011), beneficiary assessment in the World Bank, and more radically "participatory action research."

Recipients of aid and development are awash in the language and legitimating promise of participation. Here participation comes to include aspects of research, knowledge production, and evaluation most explicitly and formally. The list of cases of practical and theoretical inquiry into participation extends even further—in science, in education, in regulation, in workplaces and organizations, and so on.

Concept

If the word is rarely remarked upon, then the concept is even less well understood. Philologically speaking, it has two roots whose overlap and connection are hazy.

Methexis

The first goes back to Parmenides, and to the Greek term *methexis* (μεθεξις), which is also at the heart of Plato's theory of forms. To participate is to be self-predicating (as analytic philosophers like to say): to be an instance of something, not a copy or a representation (see **mirror**). A beautiful thing participates in the idea of Beauty; something is red because it participates in the idea Red; a given internet meme participates in the idea Internet Meme. This usage is elaborated by the Neoplatonists (Proclus, Plotinus, Pseudo-Dionysius), who articulate a tripartite relation of participant, participated, and unparticipable. The concept is worked over throughout medieval and scholastic philosophy and theology, especially in Thomas Aquinas's work, primarily as a question of the being and essence of God and God's action in the world; and to a lesser extent in Calvin's theology as a problem of the believer's participation in God through prayer. It slowly diminishes in centrality as modern philosophy progresses through the "classical age of representation," but is never abandoned. Indeed, the nineteenth century sees a "rediscovery" of the concept among the neo-Thomists of the time (McCool 1994).

A late exemplar of this conceptual use of the term is Malebranche, whose attempt to synthesize theology and rationalism, contra Descartes's solution to the problem of knowledge, directly connects the older tradition to modern philosophy. Malebranche's occasionalism investigates the problem of causality as the problem of the participation of God in the world of events. The question of causation is subsequently the philosophical battleground for the concept of participation—and one could perhaps be forgiven for drawing a connection all the way forward to the twentieth-century conundrums of quantum physics, where the participation of the observer in what she observes is debated. Indeed, the general concern with the effect of the observer on what is observed, from physics to the social sciences, is a legacy of this unfinished business with participation and causality, but the language of classical participation is not used in that debate—with the curious exception of physicist John Wheeler's concept of a "participatory universe" (Kaiser 2011; Wheeler and Patton 1975).

Finally, one last case of *methexis* that has bedeviled modern scholarship is the philosopher-anthropologist Lucien Levy-Bruhl. Levy-Bruhl used "mystical participation" to account for the difference between modern rationalist theories of causation and representation and those of "primitive mentalities" (Keck 2008). For most anthropologists, Levy-Bruhl, along with the theory of participation he espouses, is an embarrassment, of sorts; but the theory does have its defenders and considered reappropriations, from Evans-Pritchard's "test" of the theory in his work on the Azande, to Stanley Tambiah's adoption of the distinction between causality and participation (Evans-Pritchard 1937; Tambiah 1990), to a recent invocation by Marshall Sahlins in an attempt to explain kinship as "mutuality of being" (Sahlins 2013). In philosophy and sociology of science, the concept is clearly explored by Polanyi in his theories of personal knowledge and the tacit dimension of scientific practice—scientists participate in science; they do not represent it (Polanyi 1958).

Political Participation

The second aspect of the concept of participation is the more recognizable one. It is the one routinely subordinated to the concept of democracy: the relation between individuals and collectives. To participate in an election, to participate in a community, to become part of a governing body, and the like. This aspect comes with all the connotations of agency, autonomy, direct involvement, and the priority of choosing. The confusion with democratic participation is based on the common assumption that there are two forms of participation under democracy, direct and representative, the former of which is rejected and abandoned after the French Revolution, only to reemerge repeatedly (e.g., in Port Huron in 1962).

Although the term (and the word itself) is not always the central one, the concept is at the heart of Rousseau's theory of the general will, Mill's understanding of liberty, Tocqueville's characterization of American political life, and the Federalist papers' anxieties concerning tyrannies of the minority and majority. Enlightenment political thought mutates or hybridizes the theological aspect of the concept, sometimes retaining its ancient aspect (in Rousseau), sometimes abandoning it (in Mill).

In the twentieth century, there is a minor tradition of political theory, starting with Carole Pateman, that has argued for the distinctiveness of participation (Pateman 1976; Bachrach and Botwinick 1992; Mansbridge 1980). Often the concept is synonymous with small-scale, local, community organization, as in the case of Archon Fung and Erik Olin Wright's "empowered participation" in cases such as local schooling or community policing, or in a high-profile case, "participatory budgeting" (Fung and Wright 2003; Wampler 2012).

Beginning with the coining of "participatory democracy" the question was put explicitly: what is a "nonparticipatory" democracy? Why was it necessary to redouble and emphasize the "participatory" aspect of a system already assumed to rest on participation? But for mid-twentieth-century political theorists like Joseph Schumpeter, there was only one kind of participation by the people: voting. Politics for Schumpeter was simply the "competition amongst elites for votes." This "elite democratic" theory, according to people like Pateman, abandoned the classical tradition's emphasis on participation other than voting—participation in the public sphere, participation in deliberation, critique, and protest—for ideological reasons (a sociologically diagnosed "apathy" and the general suspicion about the ignorance of the people).

A "diarchic" theory of representative democracy, however, makes space for participation either through participation in the "will" of the people—elections and voting (where it is the subject of extensive research and intervention)—or as participation in free discussion in the public sphere: sometimes "deliberative democracy," sometimes public spheres or "publics and counterpublics." Under the theory of representative constitutional democracy, procedure and law circumscribe participation in order to both check and potentially guide officials in the government.

Such participation is not "direct democracy" however, because such a demand encounters a different problem: that of scale. Direct democracy is universally rejected as possible only at the small scale—the agora and the town hall where consensus decision making takes place. At any larger scale, it becomes literally impossible to incorporate the opinions and desires of everyone; representation is the only solution for modern large-scale democracies.

However, the actually existing schemes to create and implement participation under the label *participatory democracy* are rarely about legislation—they are more frequently about *administration*. They concern the carrying out of various aspects of government administration (or, in the case of worker participation, the organization and execution of tasks or jobs).

What's more, these cases of participatory administration are increasingly focused not only on deliberation and decision making, but also on the production of knowledge as such. The "wisdom of crowds" and the concept of "lay expertise" are both examples of this extension of participation to knowledge-production. In one sense this is a clear reaction to the administrative and bureaucratic reliance on expertise—with its history from the scientific management and Great Society debates of the early twentieth century. In another, it is also an extension of knowledge making to include ever-wider circles of expertise—as in "Mode 2" science, for instance (Nowotny, Scott, and Gibbons 2001). For some, the implication of this extension is troubling because it extends the function of democracy beyond the play of *doxa* (opinion) into the domain of *epistemè* (truth) (Urbinati 2014).

These two aspects of the concept combine in the present to form a kind of schizoid concept. On the one hand, it wants to signal the concerns with agency, autonomy, decision making, and involvement that are most central to the second sense described above: voice, agenda setting, direct democracy, deliberation, action. On the other hand, it also wants to signal the primary meaning of the term: to become-collective, to become an instance of a collective, not just one individual among others, but the very thing itself: social, collective, to share in the existence of many others. To belong.

Outside and beyond Participation

Participation today encounters a range of challenges from outside or beyond it (Marres 2012). Excess participation is frequently a problem, from websites overwhelmed by traffic to DDoS attacks, to the breakdown of communication or organization in protest movements, to the surprising proliferation of reality television programs. Involuntary participation is a frequent feature of modern technical platforms—something that emerges because of the constant demand to measure and monitor participation, often in order to display those metrics back to the participants as incentive or reward.

Finally, "nonparticipation" is often a question mark (Casemajor et al. 2015). Can one simply *not participate* in something, and what would it mean to do so? Despite the seemingly obvious connection, participation is not simply about inclusion or exclusion. The decision to include or exclude is a separate one from the demands or desires to participate, and it necessarily produces a different possibility: critique, denunciation, competition, sabotage, jamming. Participation is not about direct action; it is about belonging. One can belong and not participate, but one cannot participate without belonging in some way.

What might it mean that I participate in this volume, *Digital Keywords*, for instance? In this contemporary moment, participation is almost always coupled with "openness"—a demand, an offer, an explicit structure that enables known and unknown forms of participation, as in the case of free software or Wikipedia. Shouldn't *Digital Keywords* be an open website on which "anyone" can contribute or update an entry, creating an archive of keywords argued over, updated, and extended by those who know them best, and who participate in order to establish an authoritative document? Though the editor didn't go that far, the entries in this volume were nonetheless vetted publicly on the *Culture Digitally* blog, soliciting participation in peer review, editing, conception, and critique. Participation is not always open to everyone—because not everyone belongs; participation is not inclusion. Indeed, *exclusion* also has its functions. Scholarship in science and the humanities relies on a certain kind of "constitutive closure," openness within bounds. Belonging comes with an expectation of participation, but a willingness to participate does not guarantee inclusion. So even though I offered this entry to the editor without being explicitly asked—as with many such volumes, inclusion and exclusion of contributions are often fluid and negotiated—it is possible only because I already belong, in some sense, to this closed community of scholars and writers and reviewers, and already know the format and expectations of such work. Just as contributors to Linux already know what an operating system is and does, or Wikipedia editors know what an encyclopedia entry should look like: we have all become (a) collective through participation (Kelty 2012).

I end with the starting point: it is easy to miss the centrality of participation. There is a particular monument that we—scholars and teachers, readers and students—pass by every day unnoticed: the so-called participation grade on our syllabi. For many of us, this placeholder has been copied from syllabus to syllabus for years, decades—probably copied initially from someone else's syllabus. We dutifully assign an arbitrary grade to each student in each term. For some teachers it reflects attendance and etiquette, for some discussion skills, for others a certain *je ne sais quoi*—the very thing students often resent in the grade. But it does not, I wager, genuinely reflect *participation.* The classroom—in particular, the grade—is not a site of democracy; there is an experience of "belonging" but not one voluntarily chosen; and perhaps most importantly, students are rarely *instances* of the class—the class, *as such*, seems strangely the last thing on our minds, or our syllabi. It is the students' individual performances we evaluate, not that of the class as a whole. As a result we are trapped: between a participation that values the autonomy and voice of the individual, and one that values the experience of becoming-collective, of belonging. Properly speaking, participation is both of these things.

See in this volume: community, democracy, forum, internet, mirror, sharing

See in Williams: bureaucracy, collective, democracy, development, experience, expert, individual, labour, popular, representative, work

References

Arnstein, Sherry R. 1969. "A Ladder of Citizen Participation." *Journal of the American Institute of Planners* 35(4): 216–24. doi:10.1080/01944366908977225.

Asaro, P. 2000. "Transforming Society by Transforming Technology: The Science and Politics of Participatory Design." *Accounting, Management and Information Technologies* 10(4): 257–90. doi:10.1016/S0959-8022(00)00004-7.

Bachrach, Peter, and Aryeh Botwinick. 1992. *Power and Empowerment: A Radical Theory of Participatory Democracy*. Philadelphia: Temple University Press.

Beierle, Thomas C., and Jerry Cayford. 2002. *Democracy in Action: Public Participation in Environmental Decisions*. Washington, DC: RFF Press and Resources for the Future.

Benkler, Yochai. 2006. *The Wealth of Networks: How Social Production Transforms Markets and Freedom*. New Haven, CT: Yale University Press.

Bishop, Claire. 2012. *Artificial Hells: Participatory Art and the Politics of Spectatorship*. London: Verso.

Boltanski, Luc, and Eve Chiapello. 2005. *The New Spirit of Capitalism*. London: Verso.

Bourriaud, Nicolas. 2002. *Relational Aesthetics*. [Dijon]: Les Presses du réel.

Bruns, Axel. 2008. *Blogs, Wikipedia, Second Life, and Beyond: From Production to Produsage*. New York: Peter Lang.

Carpentier, Nico. 2011. *Media and Participation: A Site of Ideological-Democratic Struggle*. Bristol: Intellect.

Casemajor, Nathalie, Stéphane Couture, Mauricio Delfin, Matt Goerzen, and Alessandro Delfanti. 2015. "Non-Participation in Digital Media. Toward a Framework of Mediated Political Action." *Media, Culture & Society* 37(6): 850–66.

Chambers, Robert. 2011. "History and Development of PRA Worldwide," March. http://opendocs.ids.ac.uk/opendocs/handle/123456789/284.

Chesbrough, Henry. 2003. *Open Innovation: The New Imperative for Creating and Profiting from Technology*. Boston: Harvard Business School Press.

Cole, G. 1944. *A Century of Co-Operation*. [London]: Pub. by G. Allen & Unwin Ltd. for the Co-operative Union Ltd.

Cooke, Bill, and Uma Kothari. 2001. *Participation: The New Tyranny?* London: Zed Books.

Delwiche, Aaron, and Jennifer Henderson, eds. 2013. *The Participatory Cultures Handbook*. New York: Routledge.

Derber, Milton. 1970. *The American Idea of Industrial Democracy, 1865–1965*. Urbana: University of Illinois Press.

Deuze, Mark. 2006. "Participation, Remediation, Bricolage: Considering Principal Components of a Digital Culture." *Information Society* 22(2): 63–75. doi:10.1080/01972240600567170.

Evans-Pritchard, E. 1937. *Witchcraft, Oracles and Magic among the Azande*. Oxford: Clarendon Press.

Fish, Adam. 2013. "Participatory Television: Convergence, Crowdsourcing, and Neoliberalism." *Communication, Culture, and Critique*. http://eprints.lancs.ac.uk/61554/1/Participatory_Televisionin_the_Era_of_Convergence_prepublication_final_01.pdf.

Fish, Adam, Luis Murillo, Lilly Nguyen, Aaron Panofsky, and Christopher Kelty. 2011. "Birds of the Internet: Towards a Field Guide to Participation and Governance." *Journal of Cultural Economy* 4(2): 157–87. doi:10.1080/17530350.2011.563069.

Fisher, E., R. L. Mahajan, and C. Mitcham. 2006. "Midstream Modulation of Technology: Governance from Within." *Bulletin of Science, Technology & Society* 26(6): 485–96. doi:10.1177/0270467606295402.

Frieling, Rudolf. 2008. *The Art of Participation: 1950 to Now*. San Francisco: San Francisco Museum of Modern Art, in Association with Thames & Hudson.

Fung, A., and E. O. Wright. 2003. *Deepening Democracy: Institutional Innovations in Empowered Participatory Governance*. The Real Utopias Project Series. New York: Verso.

Gladwell, Malcolm. 2010. "Small Change: Why the Revolution Will Not Be Tweeted." *New Yorker*. http://www.newyorker.com/reporting/2010/10/04/101004fa_fact_gladwell.

Haar, Charles. 1975. *Between the Idea and the Reality: A Study in the Origin, Fate, and Legacy of the Model Cities Program*. Boston: Little Brown.

Hayden, Tom. 1962. "The Port Huron Statement." In *Takin' It To The Streets*, edited by Alexander Bloom and Wini Breines, 50–61. New York: Oxford University Press, 1995.

Hippel, Eric von. 2005. *Democratizing Innovation*. Cambridge, MA: MIT Press.

Irwin, Alan, and Brian Wynne. 1996. *Misunderstanding Science: The Public Reconstruction of Science and Technology*. Cambridge: Cambridge University Press.

Jasanoff, Sheila. 2003. "Technologies of Humility: Citizen Participation in Governing Science." *Minerva* 41(3): 223–44. doi:10.1023/A:1025557512320.

Jenkins, Henry. 1992. *Textual Poachers: Television Fans and Participatory Culture*. New York: Routledge.

Jenkins, Henry, Ravi Purushotma, Margaret Weigel, and Alice J. Robison. 2007. *Confronting the Challenges of Participatory Culture: Media Education for the 21st Century*. Chicago. https://mitpress.mit.edu/sites/default/files/titles/free_download/9780262513623_Confronting_the_Challenges.pdf.

Kaiser, David. 2011. *How the Hippies Saved Physics: Science, Counterculture, and the Quantum Revival*. New York: W. W. Norton.

Keck, Frédéric. 2008. *Lucien Lévy-Bruhl: Entre philosophie et anthropologie, contradiction et participation*. Paris: CNRS éditions.

Kelty, Christopher M. 2012. "This Is Not an Article: Model Organism Newsletters and the Question of 'Open Science.'" *BioSocieties* 7(2): 140–68. doi:10.1057/biosoc.2012.8.

Lezaun, Javier. 2011. "Offshore Democracy: Launch and Landfall of a Socio-Technical Experiment." *Economy and Society* 40(4): 553–81. doi:10.1080/03085147.2011.602296.

Lichtenstein, Nelson. 1993. *Industrial Democracy in America: The Ambiguous Promise*. [Washington, DC]: Woodrow Wilson Center Press; Cambridge: Cambridge University Press.

Mansbridge, Jane. 1980. *Beyond Adversary Democracy*. New York: Basic Books.

Marres, Noortje. 2012. *Material Participation: Technology, the Environment and Everyday Publics*. Houndmills, Basingstoke, Hampshire: Palgrave Macmillan.

McCool, Gerald. 1994. *The Neo-Thomists*. Milwaukee: Marquette University Press; Association of Jesuit University Presses.

Miller, Peter, and Nikolas Rose. 1988. "The Tavistock Programme: The Government of Subjectivity and Social Life." *Sociology* 22(2): 171–92. doi:10.1177/0038038588022002002.

Musil, Robert. 1987. *Posthumous Papers of a Living Author*. Hygiene, CO: Eridanos Press.

Nowotny, Helga, Peter Scott, and Michael Gibbons. 2001. *Re-Thinking Science: Knowledge and the Public in an Age of Uncertainty*. London: Polity Press.

Pateman, Carole. 1976. *Participation and Democratic Theory*. Cambridge: Cambridge University Press.

Polanyi, Michael. 1958. *Personal Knowledge towards a Post-Critical Philosophy*. Chicago: University of Chicago Press.

Raymond, E. 1999. *The Cathedral and the Bazaar: Musings on Linux and Open Source by an Accidental Revolutionary*. Sebastopol, CA: O'Reilly and Associates.

Rowe, G. 2005. "A Typology of Public Engagement Mechanisms." *Science, Technology & Human Values* 30(2): 251–90. doi:10.1177/0162243904271724.

Sahlins, Marshall. 2013. *What Kinship Is—and Is Not*. Chicago: University of Chicago Press.

Tambiah, Stanley. 1990. *Magic, Science, Religion, and the Scope of Rationality*. Cambridge: Cambridge University Press.

Tkacz, Nathaniel. 2015. *Wikipedia and the Politics of Openness*. Chicago: University of Chicago Press.

Turner, Fred. 2013. *The Democratic Surround: Multimedia and American Liberalism from World War II to the Psychedelic Sixties*. Chicago: University of Chicago Press.

Urbinati, Nadia. 2014. *Democracy Disfigured: Opinion, Truth, and the People*. Cambridge, MA: Harvard University Press.

Vetter, Jeremy. 2011. "Introduction: Lay Participation in the History of Scientific Observation." *Science in Context* 24(02): 127–41. http://journals.cambridge.org/abstract_S0269889711000032.

Wampler, Brian. 2012. "When Does Participatory Democracy Deepen the Quality of Democracy? Lessons from Brazil." *Comparative Politics* 41(1) : 61–81. http://www.jstor.org/stable/info/20434105.

Wheeler, John Archibald, and C. M. Patton. 1975. "Is Physics Legislated by Cosmogony?" In *Quantum Gravity: An Oxford Symposium*, edited by C. J. Isham, R. Penrose, and D. W. Sciama, 538–605. Oxford: Clarendon Press.

Wiggins, Andrea, and Kevin Crowston. 2011. "From Conservation to Crowdsourcing: A Typology of Citizen Science." In *2011 44th Hawaii International Conference on System Sciences*, 1–10. IEEE. doi:10.1109/HICSS.2011.207.

Wynne, Brian. 2007. "Public Participation in Science and Technology: Performing and Obscuring a Political-Conceptual Category Mistake." *East Asian Science, Technology and Society: An International Journal* 1(1): 99–110. doi:10.1007/s12280-007-9004-7.

22

Personalization

Stephanie Ricker Schulte

New media technologies keep coming—biometric mobile telephones, geotracking, fitness monitors, ovulation predictors—and along with them more avenues through which to link the personal and the technological. *Personalization* is, in many ways, a keyword crucial to understanding the contemporary historical and cultural moment. Personalization is enabled by technologies, embedded in infrastructures, valued by enthusiasts, targeted by discontents, coveted by marketers, and central to contemporary debates about late capitalism.

This entry is a genealogy of personalization in relation to its technological and ideological iterations. First, it discusses the origins of the term *personalization* in a variety of disciplines, moments in history, and realms of human interaction. Second, it historicizes several interrelated technological developments, including digitization, miniaturization, interactivity, and customization. Third, this entry explicates some of the key differences between interactivity, customization, and personalization to illustrate the ways ideologies and daily practices are mapped and embedded into personal technology as they are created and as human and institutional interactions shape the meanings of new gadgets. Ultimately, this entry shows that developers and distributors of personalized technologies have focused on the increased pleasure, autonomy, ease, and agency that presumably expand through and within the personalization of technology. However, this presumption of increased agency often neglects the ways in which individual practices and ideologies are nonetheless vulnerable to institutional uses and goals, in particular those that are compatible with the economic and cultural values of late capitalism.

Genealogy of the Personal

Use of the terms *personalize* and *personalization* has increased significantly in recent years.[1] In 2014, *Forbes* called it a "hot retail buzz word."[2] The origins of the terms (noun *personalization*, adjective *personal*, and verb *personalize*) all stem from the Latin *personalis* or *personale*, which means "pertaining to a person." In their update to Raymond Williams's famous *Keywords* book, Tony Bennett, Lawrence Grossberg, and Meaghan Morris identified *persona* as "one of the European world's most central and yet fluid terms."[3] The contemporary dynamics of digital personalization technologies reflect a longer preoccupation with the "person" and the "persona," with the uneasy distinction between the self and the other. These ideas root deeply in philosophy, psychology, and religion.[4] Central to the concept is the notion of the individual human and human individualism, a centrality reflected in everyday language: *personal* as a noun indicates a newspaper notice pertaining to an individual reflecting individual desires or needs; to *take something personally* means to interpret an action as directed at the individual self (such as to view something as a personal insult); *personal property* indicates items that legally belong to an individual (such as personal letters or a personal computer); to *personalize* means to mark something to indicate that it belongs to a particular individual (such as a personalized stamp or message) or to change it in accordance with an individual's preferences or needs (such as personalized medicine or education). As Raymond Williams noted in his entry on *personality*, the cognates of *personal* relate to individual "character" that is both "outward sign" and "yet internalized as a possession."[5] As the term becomes interwoven into digital spaces, it continues to represent the inextricability of internal and external selves, of the material and the symbolic.

Roots of the term *personal* in Western philosophy and political thought extend through to more contemporary moments, in particular in the ways the "personal" stands as in binary opposition to "the public." In this sense, the personal is that which is not *in* the public or *known to* the public, suggesting a kinship to what is "private." As Jeff Weintraub wrote in *Public and Private in Thought and Practice*, "the contrast between the 'personal,' emotionally intense, and intimate domain of family, friendship, and the primary group and the impersonal, severely instrumental domain of the market and formal institutions" is "one of the great divides of modern life."[6] This version of the personal—which Weintraub notes is "socio-historically variable"—indicates that which is outside the market or the political.[7] Though the keyword historically signaled concerns separated from "public" matters, post-1968 identity politics articulated a politics of personalization perhaps best encapsulated by second-wave feminism's famous phrase "The personal is political." More recently, political science has focused on the "personalization of politics," which Lance Bennett describes as an era in which "personal action frames displace collective action frames in many protest causes."[8]

Technology: Personal Computers, Interactivity, and Mass Customization

Increasingly, technology plays key roles in the meanings associated with *personal* and *personalization*. For example, when the "personal computer" consumer market emerged in the 1970s and 1980s, computers became personal property, meaning they were small and affordable enough for individuals to purchase and operate. This significant shift loosened the hold on computing technology by the large institutions and government agencies with the resources to purchase, house, and maintain, the equipment. So the "personal computer" represented a market shift. But the "personal computer" also represented a conceptual shift.[9] It represented a radical departure from "mainframe" computing, which conveyed an institutional ownership and utility. In an interview published in the *New York Times* in 1962, a computer scientist described what he called a "personal computer," a portable and affordable machine that "everyone" would have in the future.[10] These machines would identify specific users and provide individualized information, such as offering a housewife her grocery list and bank account information in the grocery store or allowing a bartender to forecast an order.

As this example illustrates, even before the technology was available, the *personal* in *personal computer* was about more than just ownership. It indicated industry and technological innovations that would put computers within reach of consumers and would make those machines reactive to individual needs as well as able to accommodate differences among individual users. *Personal* also indicated conceptual and practical innovations that would reframe computer utility in the minds and lives of individuals, reshaping what *computing* meant to include ownership and access, reactivity and accommodation, thoughts and practices, centrality and embeddedness in everyday life. The "personal" had to be conceptually connected to computing before the "personal computer" could become such. Although clearly the technology has changed, elements crucial to the "personal" in personal computing have persisted, taking the ideas rooted in the personal computer—including personal ownership, personal responsiveness, and embedment in personal home and work life—but amplifying and expanding them through much of the Western world.

With this expansion, industrial and technological innovations enabled personalization through digitization, interactivity, customization, and individualization, coconstitutive trends that collapsed boundaries between personal practices and technological devices. Digitization, or the transformation of information into discrete units, allowed information to more easily flow between devices, to be used in more varied ways and applied using new metrics. For example, in the mid-twentieth century, military, academic, and commercial institutions began creating mechanisms to convert analog sound waves into digital data sets, units of binary code that allowed sound recording on digital tape and, eventually, CDs and MP3.[11] Technologies of digitization also transformed medical records, bank accounts, and travel documents, not to mention creating individual means to map and navigate environments. These are some of the first ways the personal went digital and the digital enabled the personal.

As information became increasingly digital, technologies that processed information became portable and interactive in new ways. Computing capacity has roughly followed Moore's Law, meaning it has simultaneously become more powerful and efficient as it became cheaper, smaller. By the 1990s, people could take the descendants of personal computers with them in many of the ways imagined in the 1962 *New York Times* article. People could interact with technology in more varied and numerous ways. Indeed, interactivity became programmed and built into computers and telephones, marketed as a desirable quality in both software and hardware. As technology became reactive to user practices, behaviors, desires, and movements, devices became partners in the exchange and production of information. It became personal in a different sense. Technologies became agents with which people could interact, and even seemed to have personalities.[12] With the emergence of programs such as Clippy, the helpful aid in 1990s Microsoft Office Word, users could customize adaptive features. Users could choose wallpapers and specify ringtones for different callers. As Tom Streeter writes in his **internet** entry in this volume and in his previous work, public expansion of the internet in the late 1990s rapidly expanded the avenues for inserting unique, personal content into media venues.[13] Home page template makers and eventually social media sites lay in wait for users to interact with them, to customize online spaces with their own content and relationships. In addition to producing personalized content online, consumers also have access to increasingly customized products. Mass customization, or the mass production of customized products like children's books or running shoes, allows large companies to meet individual consumers' needs in more specific ways.[14] As consumer markets for technologies like 3D printers expand, they expand the capacities for individuals to create and more diversely tailor materials in their lives, to manufacture their own products, to conceptualize and produce their material worlds in new ways. Commodity production, like information production, is increasingly interactive and customizable.

Ideology and Practice: Customization vs. Personalization

One driving goal in creating smaller, cheaper, portable, interactive, and customizable technology was to encourage users to form affective connections with technologies. Apple fantasized users would take their iPhones "personally," in that iPhones would become ubiquitous parts of their lives. For many, this has come true. In 2013 Americans spent an average thirty-four hours a week using their smartphones.[15] Industry studies, advertising campaigns, celebratory scholarship, and to some extent policy documents[16] stress the ways interactivity and customization allow technology to meet human needs, serve people, and enable them to develop or self-actualize.[17] Human agency sits centrally to many of these hopeful narratives, which imagine technology crowdsourcing diverse cultures, politics, and products,[18] reclaiming resources previously wasted, including human attention,[19] allowing individuals to discover their authentic selves[20] or evolve beyond their bodies by merging with technology.[21] Indeed, in 2006, *Time* magazine bestowed on "You" the coveted person-of-the-year designation. The title featured a mirror and stated "You—Yes, You—Are TIME's Person of the Year," because, as the cover article noted, the "World Wide Web became a tool for bringing together the small contributions of millions of people and making them matter." As this magazine indicates, many users, journalists, and industry executives worshipped at the church of Apple, *Wired* magazine, and *BoingBoing*, seeing them as keys to humanity's salvation.

As interactive and customized technologies like mobile telephones and body monitors become more continuously used, they also produce more complex data sets about users and their social networks. The AT&T smartphone "Find My Family" app tracks individual family members using GPS in their phones. This type of data aggregated with others, such as social networking connections, can provide a "social graph," a map of the interactions between people, a map of never-before-aggregated information about human relationships and networks. This rapidly expands the boundaries of digitization, distilling what were once amorphous personal connections, massifying them, and commodifying them. This information is valuable to a number of institutions and organizations. As Deleuze has written, "individuals have become 'dividuals,' and masses, samples, data, markets or 'banks.'"[22]

The development and security of data sets—both big and small—have been the source of most of the critiques of the technology in the contemporary era. Early identity theft and hacking anxieties were followed by scandals about credit card security, Facebook and Google divulging personal information, and NSA tracking. These controversies sparked public debates that created strange bedfellows, aligning those fearing Big Brother (libertarians) on the same side as those contesting free market capitalism (lefties). In scholarly circles, the specters of Michel Foucault's surveillance and discipline expanded at the same time that Karl Marx regained popularity, especially in areas focused on the application of individual monitoring to labor.[23] Many of these well-formulated and important critiques took up the security, utility, value, or ethics of data sets and corporate/government surveillance, although few adequately theorized personalization itself as animating data sets' creation.[24] And even fewer describe with specificity—as Tarleton Gillespie does in his **algorithm** entry in this volume—the mechanisms through which the surveillance and compilation occur.

Tracking and recording unique data sets for individual people and their social networks allows marketers to provide tailored messages to unique individuals based on activities, locations, or enacted preferences. This "predictive personalization" (sometimes called hyper-personalization) uses contextual computing to divine relevant and pleasing messages, to predict what consumers will want, and to decrease advertising noise and irrelevant content. This hyper-personalization would not be possible without the technological shifts toward digitization, interactivity, and customization detailed above. A remote control and a multitude of channels—the television model of the previous generation—allowed users to customize their viewing experience by watching channels that interested them or "zapping" instead of continuously viewing. In contrast, predictive personalization allowed TiVo (or, more recently, Netflix and Amazon) to anticipate the behaviors of a consumer to facilitate an already-generated and curated personalized viewing environment.

As this example illustrates, "customization" and "personalization" are closely related but distinct concepts. While both imply creating individual-level distinctions or difference, they differ in their agency. Customization is reactive; personalization is predictive. Customization presumes the consumer choses from an array of options; personalization suggests content can be predetermined, often by algorithms in service of corporations. Customization is *by* me. Personalization is *for* me. The processes of big data and algorithmic processing allow companies to move past "mass-customization" and move toward "mass-personalization." As Gillespie notes, people often think of algorithms as autonomous and impersonal, but they are in actuality the material process of personalization, the codification of assumptions about persons, the values held by persons, into computing systems of quantitative processing. Personalizing requires a process that is imagined as deeply impersonal (data collection, aggregation, algorithmic processing). Personalizing requires the quantification of communities (Rosemary Avance: **community**), culture (Ted Striphas: **culture**), sharing economies (Nicholas John: **sharing**), the creation of surrogates (Jeffrey Drouin: **surrogate**) and analogs (Jonathan Sterne: **analog**) of relationships. Although the temptation may be to think of these terms as unrelated or as opposites, they are in fact implicated in the same processes and value sets, driven by similar structural practices, institutional actors, and corporate fantasies. In this light, the key question then becomes this: Personalization by whom, through what means, and for what purposes?

Implications

The roots of *personalization* as a digital keyword trace back to conflicted conceptualizations of the "personal." They trace back to assumptions that technologies, media content, and habits are reflections of the self or in service of the self. Indeed, technological trends toward personalization and toward the use of personal data are built on this assumption, on the assumption that data can not only accurately reflect a person, but also provide knowledge that will allow for the "right" choices to be made on behalf of that person. However, the roots also suggest ownership of the "personal." What happens when one no longer owns personal data? Is it still "personal" as such? Can it be used to "personalize" if it is no longer "personal"?

In this way, the "personal" twists back on itself. Predictive personalization requires "personal" knowledge of a person; yet "personal" has historically signified that which is not, should not, or cannot be known by others. Thus the process of personalization begs questions about connections and disconnections between the core self and the externalized self, as well as questions about what

Deleuze might call the authentic self or the assemblage.[25] How do representations of the self that are made available by a person and his or her behaviors relate to an internal self? How do strategically developed means to measure the self relate to core, performed, multiple, flexible, historically contingent, or decentered selves? To what extent are these what Foucault would call "technologies of the self"?[26] Which parts of the person are codified and reflected, and which are not, if corporate incentives or state interests are driving the collection and application of personalization? Genealogical and linguistic questions become ideological in the interaction between users and technology.

It is the inflection of the self in and through technology that makes personalization trends problematic for many scholars. Eli Pariser has noted how the filtering practices that curate media content also ensconce individuals within their preferences in what he calls "filter bubbles."[27] These bubbles not only prevent users from encountering oppositional viewpoints but also obscure decisions made in the background about how information is filtered. Cass Sunstein's work illustrates the way this runs the risk of fostering individualism, diminishing collectivity, and polarizing politics. As technologist and critic Jaron Lanier noted, technology—and the companies that produce and distribute its products—guide us into a comfortable groove.[28] Although many of these writers would not necessarily dispute the truth in the celebratory industry narratives, these and others fear that an additional effect of technological personalization is to control or shape people, to facilitate corporate aims and profits untethered to social obligations.

But to remark on the power of personalization to individualize and, potentially, isolate, polarize, or observe and control is not to brand the individual utterly powerless. Human practices first and foremost produce, regulate, make meaningful, and potentially regulate the contexts in which personal technology are developed and used. Indeed, personalization of technologies allows users to curate their own unique content, to cultivate new selves and collectivities, to expand the reach of their personal information and personal performances, to produce themselves as brands and marketers of those brands and achieve economic and social influence. They can "self-brand."[29] To be sure, this is agency, but an ambivalent agency that mixes technological capabilities and human practices within the contexts of late-capitalist, flexible economy and in increasingly competitive attention economies. Social filtering—"the selective engagement with people, communication and other information as a result of the recommendations of others"—uses a person's social network to curate information, a practice that mixes individual desires and behaviors with the goals of marketers; it mixes the strategies of technology producers with the idiosyncrasies of algorithms.[30] Social filtering hangs ambivalently, both in and outside the reach of individual agency.

Personalization helps change the terms of politics in a similar way. It enables foremost appeals to consumers through markets, not citizens through public spheres. But what happens through technology can also mobilize many of the affiliations that inform our political identifications and reconfigurations. It is political, or perhaps rather what Lauren Berlant calls "juxtapolitical" in that it feels political and radical even if it is not.[31] Indeed, by enabling personalized politics, it participates in the demise of collective action. Personalization as an organizing logic for technology also may enable the marginalization of particular political voices. After all, accusing someone of "taking something personally" has a long history as a mechanism for dismissing the insights of those experiencing injustice or structurally marginalized.

In conclusion, preexisting ideas about the personal—rooted in the term's Latin origins—animated technological development, helping drive the production and celebration of technologies as "personalized." Once technology seized the language of the personal, it transformed our relationship to the term. Indeed, personalization—as a technological trend and a set of practices—reflects a set of values, ideologies that are deeply historically, culturally, and economically situated and contingent. While the personal computer is in one sense an origin story, in another it is a story of continuity, one that reveals how technologies tap already latent values.[32] The ideologies of personalization were facilitated through technology, but, as James Carey might argue, those ideologies should not be mistaken for originating or only existing within technology.[33] Technologies seize on culture, manifest in culture, and interact with various ideas in the cultural sphere. For example, the reconstitution of values around privacy and the rise of surveillance that have emerged alongside the personalization of technology operate well within the cultural and economic contexts of late capitalism. As Adam Fish notes in his **mirror** entry in this volume, personalization may be a mirror of late capitalism, or the two may operate a hall of mirrors as the two cocreate each other in acts of reflection.

Privacy has become disincentivized, relatively difficult for individuals to maintain, and not particularly lucrative for corporations to protect. Privacy is increasingly structurally and culturally devalued, a price paid for access to media platforms, or even legally unprotected in the political sphere. But as "personalization" processes and practices accelerate and, presumably, the associated values become more unquestioned and dominant, what options for resistance remain? And what might resistance to personalization look like? "*De*-personalization?" Does media-refusal become a political position, or one embedded in identity politics?[34]

To critique personalization as a term, a technology trend, and an ideology is not to condemn it as inherently atomizing, asocial, or apolitical. It is to take aim at some of its implications in terms of the way it shifts responsibility for privacy protection to the individual against the corporation and the government. It is to outline the ways that it further builds power for corporations to collect data that could have previously been collected only by the state. A state is nominally answerable to citizens and a constitution; a state has a mandate to promote the public good. A corporation has shareholders across a multinational context; a corporation has a mandate to promote a market good. And, yes, technology does some of what it explicitly promises in the marketplace. It empowers consumers and transforms their everyday practice with personalized features, information, and networks. Personalization may determine whether we get the internet we were promised—one that connects users to one another through interactive platforms, that democratizes information, enables entrepreneurialism and civic engagement—or a different internet, which may do the above but also commercializes culture and politics, alleviates productive discomfort, facilitates surveillance, and resituates or eradicates forms of agency.

See in this volume: algorithm, analog, community, culture, internet, mirror, sharing, surrogate

See in Williams: capitalism, community, consumer, democracy, exploitation, individual, media, ordinary, personality, private

Notes

Thank you to Tarleton Gillespie, Ben Peters, Gabriella Coleman, Fred Turner, Jonathan Sterne, Laura Cook Kenna, Julie Passanante Elman, and several anonymous Digital Keywords participants for their feedback.

1 An increase of 714 percent between the 1960s and 2000s (Collins Dictionary Trend Metric); 417 percent increase between 2005 and 2014 (Google News Headline Metric).

2 Barbara Thau, "How Big Data Helps Stores Like Macy's and Kohl's Track You Like Never Before," *Forbes*, January 24, 2014: http://www.forbes.com/sites/barbarathau/2014/01/24/why-the-smart-use-of-big-data-will-transform-the-retail-industry/.

3 Tony Bennett, Lawrence Grossberg, and Meaghan Morris, *New Keywords: A Revised Vocabulary of Culture and Society* (Malden, MA: Blackwell Publishing, 1988), 254–55.

4 For philosophy, see Aristotelian and Platonic debates about the limits of personal perspective and Cartesian dualism mind-body debates; for psychology, see Jungian personal unconscious and Freudian psychoanalysis; and for religion, see Martin Luther's belief in a personal relationship with God.

5 Raymond Williams, *Keywords: A Vocabulary of Culture and Society*, rev. ed. (New York: Oxford University Press, 1983), 232–35.

6 Jeff Weintraub and Krishan Kumar, *Public and Private in Thought and Practice: Perspectives on a Grand Dichotomy* (Chicago: University of Chicago Press, 1997), 20.

7 Ibid., 37.

8 Lance Bennett, "The Personalization of Politics: Political Identity, Social Media, and Changing Patterns of Participation," *Annals of the American Academy of Political and Social Science* 644 (2012): 20–39.

9 Bryan Pfaffenberger, "The Social Meaning of the Personal Computer: Or, Why the Personal Computer Revolution Was No Revolution," *Anthropological Quarterly* 61(1) (1988): 39–47.

10 "Pocket Computer May Reduce Shopping List," *New York Times*, November 2, 1962.

11 Debates persist about the relative quality of these sounds, but these technologies allowed sound to be more easily transported among platforms, paving the way for flexible, individual soundscapes enabled by Walkmans, iPods, and cellular ringtones. Todd Gitlin, *Media Unlimited: How the Torrent of Images and Sounds Overwhelms Our Lives*, rev. ed. (New York: Picador, 2007).

12 Thanks to Tarleton Gillespie for making this connection.
13 Thomas Streeter, *The Net Effect: Romanticism, Capitalism, and the Internet* (New York: New York University Press, 2011).
14 Rebecca Bagley, "How 3D Printing Can Transform Your Business," *Forbes*, May 3, 2014, http://www.forbes.com/sites/rebeccabagley/2014/05/03/how-3d-printing-can-transform-your-business/.
15 "How Smartphones Are Changing Consumers' Daily Routines around the Globe," *Nielsen*, February 24, 2014, http://www.nielsen.com/us/en/newswire/2014/how-smartphones-are-changing-consumers-daily-routines-around-the-globe.html.
16 Such as policy documents that fund broadband access.
17 Stephanie Schulte, *Cached: Decoding the Internet in Global Popular Culture* (New York: New York University Press, 2013).
18 Henry Jenkins, Sam Ford, and Joshua Green, *Spreadable Media: Creating Value and Meaning in a Networked Culture* (New York: New York University Press, 2013); Eric Schmidt and Jared Cohen, *The New Digital Age: Reshaping the Future of People, Nations and Business* (New York: Knopf, 2013).
19 Clay Shirky, *Cognitive Surplus* (New York: Penguin Press, 2010); Jane McGonigal, *Reality Is Broken: Why Games Make Us Better and How They Can Change the World* (New York: Penguin, 2011).
20 Kevin Kelly, *What Technology Wants* (New York: Viking Books, 2010).
21 Ray Kurzweil, *The Singularity Is Near* (New York: Viking Books, 2005).
22 Gilles Deleuze, "Postscript on the Societies of Control," *October* 59 (1992): 5.
23 For Foucault, see Nikolas Rose, *Governing the Soul: The Shaping of the Private Self* (London: Free Association Books, 1999). For Marx, see Andrew Ross, "In Search of the Lost Paycheck," and Mark Andrejevic, "Estranged Labor," in *Digital Labor: The Internet as Playground and Factory*, ed. Trebor Scholz (New York: Routledge, 2013).
24 Kelly Gates, *Our Biometric Future: Facial Recognition Technology and the Culture of Surveillance* (New York: New York University Press, 2011); Joseph Turow, *Daily You: How the New Advertising Industry Is Defining Your Identity and Your Worth* (New Haven, CT: Yale University Press, 2011); Evgeny Morozov, *To Save Everything, Click Here: The Folly of Technological Solutionism* (New York: PublicAffairs, 2013).
25 Gilles Deleuze and Félix Guattari, *A Thousand Plateaus: Capitalism and Schizophrenia* (Minneapolis: University of Minnesota Press, 1987).
26 Michel Foucault, *Technologies of the Self: A Seminar with Michel Foucault* (Amherst: University of Massachusetts Press, 1988).
27 Eli Pariser, *The Filter Bubble: How the New Personalized Web Is Changing What We Read and How We Think* (New York: Penguin Books, 2012).
28 Cass Sunstein, *Republic.com 2.0* (Princeton, PA: Princeton University Press, 2001); Jaron Lanier, *You Are Not a Gadget* (New York: Alfred Knopf, 2010).
29 Sarah Banet-Weiser, *Authentic™: The Politics of Ambivalence in a Brand Culture* (New York: New York University Press, 2013); Alison Hearn, "'Meat, Mask, Burden' Probing the Contours of the Branded 'Self,'" *Journal of Consumer Culture* 8(2) (2008): 197–217.

30 Michele Willson, "The Politics of Social Filtering," *Convergence: The Journal of Research into New Media Technologies* 20(2) (2014): 218–32.
31 Lauren Berlant, *The Female Complaint* (Durham, NC: Duke University Press, 2008).
32 Anna McCarthy, *Citizen Machine: Governing by Television in 1950s America* (New York: New Press, 2010). This book illustrates this in advertising in an earlier era.
33 James Carey, *Communication as Culture* (New York: Routledge, 1992).
34 Laura Portwood-Stacer, "Media Refusal and Conspicuous Non-consumption: The Performative and Political Dimensions of Facebook Abstention," *New Media & Society* 10(1) (2012): 1041–57.

23

Prototype

Fred Turner

Silicon Valley is a land of prototypes. From cramped, backroom start-ups to the glass-walled cubicle farms of Apple and Oracle, engineers labor day and night to produce working models of new software and new devices on which to run it. These prototypes need not function especially well; they need function hardly at all. What they have to do is make a possible future visible. With a prototype in hand, a project ceases to be a pipe dream. It becomes something an engineer, a manager, and a marketing team can get behind.

But this is only one kind of prototype, and in many ways it's the easiest to describe. Silicon Valley produces others, sometimes alongside software and hardware, in the stories salesmen tell about their products, and sometimes well away from the digital factory floor, in the lives that engineers and their colleagues lead. When salesmen pitch a new iPhone or, say, new software for mapping your local neighborhood, they often also pitch a new vision of the social world. Their devices Will Change Human History For The Better—and you can glimpse the changes to come right there, these hucksters suggest, in the stories they tell. As they enter the marketplace, the technology-centered worlds these storytellers have talked into being become models for society at large. Likewise, when engineers and their colleagues gather at festivals like Burning Man, or even when they huddle in the tiny, underfinanced, hyperflexible teams that drive start-up development, they engage in modeling and testing new forms of social organization, often self-consciously. Like the constellations of people and machines described in marketing campaigns, these modes of gathering have technologies at their center, but they are also prototypes in their own right—of an idealized form of society.

These *social* prototypes present a puzzle for those who take *prototype* to be a digital keyword: How is it that a term so closely wedded to engineering practice should also be so clearly applicable to the nontechnical social world? Much of the answer depends on the work of hardware and software engineers, who have exported their modes of thinking and working far beyond the confines of Silicon Valley. But much also depends on the peculiarly American context in which these engineers work. In the United States, the concept of the "prototype" has a dual history. It is rooted in engineering practice, but it is also rooted in Protestant and especially Puritan theology. Few if any Silicon Valley engineers would call themselves Puritans, of course. But by briefly tracing these two traditions, I hope to show how a region long thought to depend on a uniquely Californian ideology has in fact anchored its work in some of the deepest harbors of America's capitalist mythos.[1] In the process, I hope not only to excavate the history of the term *prototype* but, through it, to begin to explain how and why Silicon Valley has itself become a model metropolis in the minds of many around the world.

The Prototype in Software Engineering

Within the world of software and computer engineering, the prototype is a relatively new arrival. In other industries, three-dimensional models of forthcoming products have been the norm for generations. Architects have long built scale models of houses, for instance, just as ship-makers have built scale models of their vessels. These models give three-dimensional life to measurements first defined on a blueprint, just as the blueprint gives two-dimensional form to ideas that emerged in conversations between the architect, the ship-maker, and their clients. For industries such as these, prototypes have long constituted an ordinary link in a chain of activities by which ideas become defined, modeled, and built.

Until the late 1980s, most software architects approached a new project simply by attempting to define its features on paper in something called a "requirements document."[2] Many still do today. One technical writer describes the process thus: "Take a 60-page requirements document. Bring 15 people into a room. Hand it out. Let them all read it."[3] This process has a number of advantages. First, such documentation produces very precise specifications in a language that all developers can understand. Second, the document can be edited as the project evolves. Third, because it lives on paper and usually in a binder somewhere in an office, the continuously updated requirements document can serve as a repository, a passive reminder of what the team has agreed to do.

Unfortunately, requirements documents can also leave developers unable to see their work whole. After handing out a large requirements document and letting everyone read it, the technical writer above says, "Now ask them what you're building. You're going to get 15 different answers." Requirements documents can confuse developers as well as inform them. They can also leave out users. Developers routinely talk with their clients before drafting requirements documentation, but they often discover that users' actual needs change as systems come online. Translating these changes into the requirements documents and then back again into the product can be complicated and time-consuming. Finally, diagrams do little to help systems developers and clients create a shared language in which to discuss these changes.[4]

Enter the prototype. In a 1990 manual for developers entitled *Prototyping*, Roland Vonk argued that building a working if buggy software system could transform the requirements definition phase of system development. The prototype could become an object, like an architect's model, around which engineers and clients could gather and through which they could articulate their needs to one another. It would speed development, improve communication, and help all parties arrive at a better definition of requirements for the system.

It would also be fun. "Prototypes encourage play," wrote one developer.[5] In the process, they also allow various stakeholders to make an emotional investment in the future suggested by the model at hand. Being by definition incomplete, prototypes encourage stakeholders to work at completing the object. Playing with prototypes helps stakeholders not only imagine, but, to a limited degree, act out the future the prototype exemplifies. The experiential aspect of prototypes also renders the projects they represent especially available to the kinds of performances and stories out of which marketing campaigns are made. Consider this brief account, penned by the designer of a computer joystick:

> Our first prototypes gave [the client firm] Novint and its investors a first peek at what was an exciting, yet nascent, concept. We started with sexy prototypes (we call them *appearance models*) that captured a vision for what the product might become down the road. By sexy, I mean models in translucent white plastic and stainless steel that took their cues from the special effects found in science fiction movies that gamers enjoy. This created a target for what the final product could be and also helped the company build investor enthusiasm around the product idea.
>
> With . . . our first prototypes in hand, Novint could create a narrative about where it was headed with this product. It was a story that now had some tangible components and emotional appeal, thanks to the physical models prototyped by [our] designers. That was a promising start.[6]

As Lucy Suchman and others have pointed out, information technologies represent "socio-material configurations, aligned into more or less durable forms."[7] Prototypes represent sites at which those configurations come into being. Prototypes simultaneously make visible technical possibilities and actively convene new constituencies. These stakeholders can help bring the technology to market, but they also represent new social possibilities in their own right. The pattern in which they've gathered can itself become a model for future gatherings, within and even beyond the industry in question.

Daniel Kreiss has put this point succinctly: "While most of the literature on prototypes focuses on small-scale artifacts and research labs, there is no theoretical reason why prototypes do not also exist at the field level."[8] Kreiss has tracked the use of what he calls "prototype campaigns" across several presidential voting cycles. In a 2013 paper for *Culture Digitally*, he explored two: the Howard Dean and Barack Obama campaigns of 2004.[9] The Dean campaign took exceptional advantage of digital technologies. It recruited leading consultants and computer scientists, built powerful databases of voters, and established a visible web presence. Dean staffers called their work an "open-source" campaign. In the process, as Kreiss explains, they not only aligned various stakeholders around computers and data; they also turned their use of computers and data into evidence that they belonged at the center of a much larger cultural story. Through that story, they claimed the kind of cultural centrality and national legitimacy that most outsider candidates can only dream of.

When the Dean campaign imploded, the Obama campaign was only too happy to adopt key members of his technology team and to claim that Obama too was running a bottom-up, technology-enabled campaign. As Kreiss has shown, they were not. On the contrary, the Obama campaign used computers to centralize and manage the same kinds of data and power on which elections have always depended.[10] But as a symbol, the Obama campaign seemed to model a world emerging simultaneously in the computer industry, a world that Americans could imagine would be open, networked, individualistic, and free.

Change by Design

There is a tension here between the sense of the campaign itself as a prototype and its depiction as a prototype. In Suchman's account, information technologies generate social arrangements. In Kreiss's, the sociotechnical arrangements of campaigns become elements of stories that in turn legitimate future actions. For the designers of the Novint joystick, prototypes play both roles. Taken together, these three accounts remind us that the material, technical, and organizational elements of prototypes are always also potentially symbolic. Advocates within an engineering firm or a political campaign can turn them into stories. Outsiders such as journalists can also take them up and turn them into the elements of national or even globe-traveling narratives. In each case, particular sociotechnical configurations become available as potential visions of a larger and presumably better way of organizing society as a whole.

Within Silicon Valley, there are a host of organizations devoted to identifying and promulgating promising social prototypes.

These include futurist outfits, research firms, and venture capitalists, among many others. Few firms transform engineering prototypes into social prototypes more self-consciously or more visibly than the Palo Alto–based design firm IDEO. Founded in 1978, the firm applies what it calls "design thinking" to every aspect of its client organizations, including individual products and brands, as well as software development, communication strategy, and organizational structure. For any given product, the firm can coordinate every aspect of the prototyping process at the engineering level; at the same time, it can link the devices and processes that emerge to new kinds of stories.

To get a feel for how IDEO transforms engineering prototypes into social prototypes, one need only consult CEO and president Tim Brown's 2009 book, *Change by Design: How Design Thinking Transforms Organizations and Inspires Innovations.* Part business how-to, part advertisement for IDEO, the book outlines the firm's philosophy of "design thinking" and shows how it has worked in a variety of specific cases. Within design thinking, prototyping occupies two places. The first would be easy for most anyone in Silicon Valley to recognize as an ordinary part of manufacturing. Prototyping stands as the opposite of "specification-led, planning-driven abstract thinking."[11] IDEO founder David Kelly calls it "thinking with your hands."[12] As Tim Brown points out, prototyping can be cheaper and faster than simply drawing diagrams, and it can engage users in shaping products as they emerge. Brown also argues that to enable prototypes to have real impact, designers need to embed them in stories. These "plausible fictions," says Brown, help designers keep their end users in mind and help potential customers, within and outside the firm, imagine what they might do with the objects and processes being prototyped.[13]

Thus far, Brown's discussion of prototypes echoes conversations in most any prototype-oriented engineering space. But toward the end of his book, Brown takes a millenarian turn. "We are in the midst of an epochal shift in the balance of power," he argues. Corporations have turned from producing goods to producing services and experiences. Customers have become something more than mere buyers. According to Brown, they have become collaborators, coconstructors of the product-experiences they acquire. Lest the reader imagine this to be a purely commercial transformation, Brown argues that "what is emerging is nothing less than a new social contract"—a contract so revolutionary that it could save the planet: "Left to its own, the vicious circle of design-manufacture-marketing-consumption will exhaust itself and Spaceship Earth will run out of fuel. With the active participation of people at every level, we may just be able to extend this journey for a while longer."[14]

The notion that consumer choice and political choice can be fused, and that, together, they can save humanity from itself, has haunted the marketing of digital media for more than twenty years. But there is more than marketing at stake in *Change by Design*. For Brown, prototyping has become a way to transform the local, everyday work of engineering into a mode of personal spiritual development. "Above all, think of life as a prototype," writes Brown:

> We can conduct experiments, make discoveries, and change our perspectives. We can look for opportunities to turn processes into projects that have tangible outcomes. We can learn how to take joy in the things we create whether they take the form of a fleeting experience or an heirloom that will last for generations. We can learn that reward comes in creation and re-creation, not just in the consumption of the world around us. Active participation in the process of creation is our right and our privilege. We can learn to measure the success of our ideas not by our bank accounts but by their impact on the world.[15]

For engineers, prototypes must be things or stories. For analysts like Suchman and Kreiss, as well as for engineers, they can be constellations of people and things that become elements in narratives that in turn have marketing or political force. But for Brown, prototyping is something much more. Prototypes as he describes them belong to a way of looking at the world in which individuals constantly remake themselves, in which they test themselves against the world and, if they find themselves wanting, improve themselves. Their quest for self-improvement in turn models the possibility of global transformation. In this vision, making a better product in the factory models and justifies the process of making a better self in everyday life. Making both together, through the process of participation and with proper attention to metrics and measurement, might even prevent the apocalyptic crash of Spaceship Earth.

Puritan Typology

Brown's world-saving rhetoric is a staple of Silicon Valley. But it did not originate there. To understand how Brown and his readers could imagine themselves as prototypes, we need to turn backward in time, trek three thousand miles to the east, and revisit the Puritans of colonial New England. When the Pilgrims landed on Cape Cod, they brought with them an extraordinarily rich practice of biblical exegesis that they called "typology." In their view, as in the view of biblical scholars all the way back to Saint Augustine, events in the Old Testament served as "types"—which we would now call "prototypes"—of events in the life of Christ recounted in the New Testament.[16] When Jonah spent three days in the belly of a whale, for example, he foreshadowed Christ's burial and resurrection.[17] For the Puritans, types were not simply symbols in stories; rather, they represented God's efforts to speak to fallen man through his limited senses. In this biblical view, Jonah really did go down under water, and when he rose up, he sent word out through time that soon Christ himself would go down under the earth and rise up too.

For the Puritans, typology did not stop at the level of the text. Rather, it offered them a vision of the world *as a text*. In the typological view, God had written his will into time. History consisted of a series of prophecies, rendered in the world as prototypical events, and fulfilled by later happenings. The biblical exodus of the Israelites, for instance, foreshadowed the migration of the Puritans themselves from England to the New World. To their congregants, the Puritan ministers of Boston and Cambridge seemed to have been prefigured by the saints of the Bible and to serve as types of saints yet to come. Each individual's life was little more than a single link in a chain of types. On the one hand, an individual such as the prolific New England Puritan minister Cotton Mather might see himself as the fulfillment of a mode of sainthood prophesied in the Bible. And on the other, his congregation might see him as an example to follow into a heavenly future. For the Puritans, history moved ever forward toward the completion of divine prophecy. But the type—or, again, prototype—pointed both forward and backward in time. The Puritan type was a hinge between past and present, mortal and divine.

For individual Puritans, the ability to read the world as a series of types carried enormous meaning. The doctrine of predestination, to which all New England Puritans subscribed, asserted that God had already decided whom to save and whom to send to hell. There was nothing anyone could do about his or her fate. As Max Weber pointed out long ago, this belief set off an extraordinary effort among living Puritans to spot signs of their possible election.[18] After all, what God could be so cruel as to curse in life those he was about to save for all eternity? By the early 1700s, the signs of likely salvation included most prominently the ability to read the natural world of New England as a series of types, written into history by God. Prototyping has long foretold brighter futures.

By now, you may have begun to wonder what, if anything, seventeenth- and early eighteenth-century theology might have to do with contemporary science and engineering. One answer is simply that Silicon Valley is suffused with the same Protestant ethic that drives other manufacturing regions. But there is another, more historically specific answer too. It was in early eighteenth-century New England that Newtonian physics met Puritan theology, and it was there that American scientists and engineers first linked scientific progress and Puritan teleology. No one did this more gracefully than the minister Jonathan Edwards. Though many remember Edwards today as the author of the quintessential fire-and-brimstone sermon "Sinners in the Hands of an Angry God," Edwards also wrote widely on science and philosophy. Throughout his life he kept a notebook in which he recorded his struggles to fuse the scientific and the divine. Published under the title *Images or Shadows of Divine Things* in 1948, the notebook simply records the types that Edwards believed he saw in nature.

Consider the following fairly representative entry:

> The whole material universe is preserved by gravity or attraction, or the mutual tendency of all bodies to each other. One

> part of the universe is hereby made beneficial to another; the beauty, harmony, and order, regular progress, life, and motion, and in short all the well-being of the whole frame depends on it. This is a type of love or charity in the spiritual world.[19]

For Edwards, gravity explicitly modeled God's love for man. But implicitly, Newton's discovery of gravity and Edwards's own ability to recognize gravity as a type marked Newton and Edwards as potential members of God's elect. In Edwards's typological history, theology and science marched hand in hand toward the end of time, each illuminating God's will and each producing saints to do that work.

Which brings us back to Tim Brown, IDEO, and Silicon Valley. For some time now, analysts have suggested that the digital utopianism that continues to permeate Northern California came to life only there. In fact, an archeological exploration of the term *prototype* reveals that the habit of linking scientific and engineering practice to a historical teleology rooted in Christian theology can be traced back to New England, if not farther. As he declaims the power of design thinking to save the world, Tim Brown echoes the Puritan divines of centuries past. They too called on their readers to see their lives as prototypes, and to see prototyping as a project that might save their souls and perhaps even the fallen world. Though Brown nowhere refers to God, his volume fairly aches with a longing to find a global meaning in his life and work, to know that he and IDEO are on the side of the angels, that they are not just fallen souls, marketing their wares as best they can, in the corrupt metropoles of capitalism.

So What *Are* Prototypes?

With this brief history of Puritan typology in hand, we can begin to complicate both the picture of prototypes that we have received from engineering and the picture of Silicon Valley that we have received from historians and marketers. In computer science and many other disciplines, engineers build prototypes to look forward in time. They hope to anticipate challenges, reveal user desires, and engage stakeholders in the kinds of experiences that will generate buzz about the product, within and beyond the boundaries of the firm. In Silicon Valley, as elsewhere, intermediaries such as IDEO turn these constellations of technologies and people into elements in stories that can in turn serve to legitimate and even model new social forms. To the extent that we see prototypes as exclusively forward looking, then the process of turning engineering and its products into models of ideal social worlds may look simply like another stage in the conquest of everyday life by the information industries of Silicon Valley.

Yet, as Puritan typology reminds us, prototypes always look backward in time as well as forward. The means by which they gather society and technology have their roots in worlds that precede and prefigure the futures they will call out for. And the particular mode of prototyping practiced by Tim Brown and many others in Silicon Valley has its roots not only in the world of engineering, but in the theology of Puritan New England. When he and others turn individual products and processes into prototypes of an ideal social world, they are following in the footsteps of Puritan divines like Jonathan Edwards. They are hardly Puritans in any theological sense. But they are not just contemporary Californians either. Like the self-proclaimed prophets of seventeenth-century New England, they are seeking to reveal a hidden order to everyday life. They too hope to uncover a hidden road to heaven and to take their place as saints along the way. They too are wondering whether they have been chosen. And they are offering prototyping to their audiences as a method by which they too might discover their own election.

The affordances of engineering prototypes assist in this process. Because prototypes are incomplete, half-cooked, in need of development, they solicit the collaboration of users and others in the building of a particular future. Because prototypes emerge from the laboratory or the office, they can seem to have no politics. They become enormously difficult to recognize as carriers of a particular teleology. On one hand, they begin to shadow forth a new social order, one in which engineers and marketers become ministers, the marketplace a kind of congregation, and Silicon Valley a new sort of city on a hill. On the other, the seeming ahistoricity of the prototype shields its makers and the breadth of their ambitions from recognition.

For all of these reasons, we need to ask new questions of the prototypes we encounter. We need to ask, How does a given prototype summon the past, as well as foreshadow a particular future? For what purposes? What sort of teleology does it invoke? And what sort of historiography does it require? How do prototypes leave the lab bench and the coder's cubicle to become elements in stories about the world as a whole? How do engineering prototypes become social prototypes? And who wins when they do?

By answering these questions, we might finally begin to stop thinking of our lives as prototypes and of new technologies as foreshadowings of a divine future.

See in this volume: algorithm, analog, archive, cloud, digital, gaming, internet, surrogate

See in Williams: art, consumer, development, image, industry, institution, myth, production, progressive, representative, technology, work

Notes

1 Richard Barbrook and Andy Cameron, "The Californian Ideology," http://www.imaginaryfutures.net/2007/04/17/the-californian-ideology-2/.
2 Roland Vonk, *Prototyping: The Effective Use of CASE Technology* (New York: Prentice Hall International, 1990), X–XI.
3 Todd Zaki Warfel, *Prototyping* (Rosenfeld Media, November 1, 2009), Safari Books Online, sec. 1.3.
4 Ibid., X.
5 Ibid., sec. 1.3
6 John Edson, *Design Like Apple: Seven Principles for Creating Insanely Great Products, Services, and Experiences* (John Wiley & Sons, July 10, 2012), Safari Books Online, sec. entitled "Prototype and the Object."
7 Lucy Suchman, Randall Trigg, and Jeanette Blomberg, "Working Artefacts: Ethnomethods of the Prototype," *British Journal of Sociology* 53(2) (June 2002): 163–79; 163.
8 Daniel Kreiss, "Political Prototypes: Why Performances and Narratives Matter," *Culture Digitally*, http://culturedigitally.org/2013/11/political-prototypes-why-performances-and-narratives-matter/, posted November 22, 2013.
9 Ibid.
10 Daniel Kreiss, *Taking Our Country Back: The Crafting of Networked Politics from Howard Dean to Barack Obama* (New York: Oxford University Press, 2012).

11 Tim Brown and Barry Katz, *Change by Design: How Design Thinking Transforms Organizations and Inspires Innovation* (New York: Harper Business, 2009), 89.
12 Kelly, quoted ibid.
13 Brown, *Change by Design*, 94.
14 Ibid., 178.
15 Ibid., 241.
16 Ursula Brumm, *American Thought and Religious Typology* (New Brunswick, NJ: Rutgers University Press, 1970), 26.
17 Perry Miller, introduction to Jonathan Edwards, *Images or Shadows of Divine Things*, ed. Perry Miller (New Haven, CT: Yale University Press, 1948), 1–42; 6.
18 Ibid., 27; Max Weber, *The Protestant Ethic and the Spirit of Capitalism* (London: Routledge 2001), 65–72.
19 Edwards, *Images or Shadows of Divine Things*, entry 79, p. 79.

24

Sharing

Nicholas A. John

Sharing, in digital contexts, can simply refer to the transfer of data from one place to another, or to making some data available to other people or machines. This is certainly how the term was used in describing the various arrangements by which data were transported between the entities and programs exposed by Edward Snowden in the summer of 2013. However, while the term *data sharing* would not appear controversial in any way, the same certainly cannot be said of *file sharing*, despite its equally deep roots in the field of computing. File sharing, assert certain representatives of the state and the entertainment industry, is not sharing, but rather theft. Critical voices of quite a different ilk might point out that Facebook, Google, and the rest do not "share" information about users with third parties, which is the language used in such companies' terms and conditions; rather, and more linguistically accurately, they sell it. Both of these objections to the use of the word *sharing*—despite their quite different political motivations—are equally revealing. What they reveal is that, for many people, sharing is a cherished notion that must not be sullied; some things may properly be described as sharing, while others most certainly may not.

The layers of meaning conveyed by the keyword *sharing* often escape our attention. According to popular wisdom, sharing is caring, and an online image search for that phrase—which produces an abundance of teddy bears and pink hearts—uncovers a deep well of cultural associations that it draws from.[1] Without these cultural associations, it is hard to account for the icon chosen by Dropbox to accompany the word *Share* on its website: while the words *File*, *Photos*, and *Links* have quite standard icons by way of illustration, the small image next to the word *Share* is a rainbow.

Sharing, according to Dropbox rhetoric, is about happiness, optimism, peace, and cooperation, and the equating of sharing with rainbows makes for an almost irresistible target: Whom do those rainbows serve? How much online sharing is dishonest and deceitful? When social network sites (SNSs) talk about sharing, to what extent are they being manipulative and exploitative? These are important questions, but they do not address the power of the word *sharing*. In particular, they do not help us understand how "sharing" became perhaps *the* overarching concept for our involvement in digital environments, nor do they help us understand how the keyword frames, and hence encourages, our involvement in them. This entry, though, seeks to do just that.

In this sense, then, *sharing* has much in common with **community** and **forum**, especially the idea that online versions are somehow a distortion or even an ideological usurpation of what these terms "really" refer to. In discussing *sharing* as a digital keyword, though, we must look beyond these criticisms of the word's use and search for a richer understanding of its workings as a metaphor that saturates our usage of information and communication technologies (ICTs) today. *Sharing* is an important digital keyword not only because of its roots in computing (time sharing, disk sharing, file sharing, etc.), but because it bears the promise that today's network and mobile technologies—because they make it easier for us and encourage us to share extensively—will bring about a better society. Given the myriad creative ways in which precisely these technologies are used to challenge the powerful and offer alternative ways of doing and being, this is a promise that should not be dismissed out of hand. At the same time, given the many ways in which businesses monetize and governments surveil how we share information digitally, nor should that promise be naively accepted.

As a keyword of the digital age, *sharing* is the constitutive activity of Web 2.0, or the interactive, user-generated internet. It is what we do on social network sites; it is the name given to the act of distributing data (photos, links, videos, tweets, etc.) through electronic networks; it is what we do when leaving a review on Amazon ("Share your thoughts," prompts the site) or uploading a movie to YouTube; it is the act of updating our status on Facebook or checking in on Swarm; all of these, and more, come under the extremely broad umbrella of *sharing*.

The word *sharing* has a history, even within the context of social network sites. To understand something of this history, I have analyzed the emergence and development of the word *sharing* in forty-four different SNSs from 1999 to 2010 (after which I have not observed significant changes).[2] The overall trend in the usage of the word is from the specific to the far more general, and the introduction of the word to describe existing activities that had previously been called something else (posting, sending, updating, etc.). We have also witnessed the introduction of what I am calling a *communicative logic of sharing*, in addition to its obvious *distributive logic*.

In the early 2000s, when we shared something on a social network site, the object of our sharing was clearly defined: we were invited to share web links and photos and the like. However, from 2007 (and not before), on their home pages SNSs started urging us to share what I call "fuzzy objects of sharing," which included things such as "your life," "your world," or "the real you." These invitations are "fuzzy" because it is not explicitly clear what it is that we are meant to share. I understand what it means to share photos or videos, but what exactly does it mean to "share your world"? When the object of sharing is fuzzy, the act of sharing online becomes far broader, the rhetorical motivation appeals more openly to an ethics of generosity, and the net of subsequent information capture is cast wider. Sharing your world implies letting other people know as much as possible about what you are doing, thinking, and, importantly, feeling. Its vagueness—or fuzziness—enables its comprehensiveness.

Another innovative use of the term *share* in SNSs, which also reflects its increasing generalization, was that it began to be used without any object at all, but simply on its own, sometimes as part of the site's self-description (in texts that might read, "On this site you can share with your friends"), and sometimes in the form of an imperative ("Share!"). This represents an important stage in the short history of the keyword *share* in the context of Web 2.0, as it indicates to us that SNS users were now assumed to know what sharing is without having to be told what to share. Significantly for a historically informed understanding of this keyword, none of the

SNSs that I analyzed used the word *share* without an object prior to 2005 (which is to suggest that if this book had been published on the thirtieth anniversary of Williams's text, the inclusion of *sharing* as a digital keyword would not have felt quite so natural as it does now, on the fortieth anniversary).

Third, in the second half of the 2010s, as *sharing* became the word to describe our participation in SNSs (and recall that Facebook opened its doors to all comers in October 2006), a number of sites started replacing words such as *update*, *post*, and *send* with *share*. The functionality of these sites had not changed, but their rhetoric did. This reflects the ascension of "sharing" to its current position as the constitutive activity of Web 2.0: not only did SNSs feel that users were now familiar enough with them that they could talk of sharing without indicating what is to be shared; many of them recognized that if they were not describing themselves in terms of sharing, they had better start to do so; they recognized that "sharing" is the name of the game, and if they wanted to position themselves within the booming SNS industry, they had better be in the sharing business. Hence, for instance, on the photo-sharing SNS Fotolog, the tag line "Make it easy for friends/family to see what's up with you" was replaced with "Share your world with the world."

Sharing, though, is not just the keyword for social network sites. It participates in a broad register of current digitally mediated communication vocabulary. Thus, for instance, mobile service provider T-Mobile ran an ad campaign with the tag line "Life's for Sharing." Again it may be tempting simply to dismiss this use of the word *sharing* as driven by an ideological profit motive: commercial enterprises seek to entice us to use their services by cloaking their gain in a language of altruism and concern for others, runs the argument, whereas "sharing" (and here the scare quotes become particularly pertinent) only serves the companies' bottom line and makes us even more narcissistic. But if we dig deeper into this keyword, we can see it as bearing a larger social promise of the digital era, as the new meanings of *sharing* bleed into older and almost exclusively positively valenced meanings of the term. Sharing, as I shall attempt to briefly show, then becomes the model for a digitally based readjustment of our interactions with things (sharing instead of owning) and with one another (sharing as the form of communication on which our relationships—especially, but not exclusively, romantic ones—are based). In order to make this point, though, we need to reach back further than the invention of SNSs.

In its nonmetaphorical sense, to share is to divide. By the plowshare the earth is rent asunder; *share* and *shear* were once the same word, with their roots in the Old English *scearu*.[3] When a child shares her candies, she divides them between her and her friends; when a child shares a toy with another, he is offering joint access to the toy; also, if I own shares in a publicly owned company, I own part of that company. Sharing, then, at least from the sixteenth century, is about distribution of scarce resources—both dividing stuff up and granting more than one person access to the same thing, while the word itself appears agnostic about the object of sharing. There are rules and norms for sharing that are not wildly divergent from those that govern gifting—there is an implicit expectation of reciprocity, for instance ("If you don't ever share your candies with me, I'll stop sharing mine with you"). As with gifting, sharing also creates and sustains social ties.[4] Some things we share actively and voluntarily; other things we share passively and by necessity—including infrastructures, public spaces, or even our planet.[5]

More recently—that is, within the last hundred years or so—in addition to referring to distribution, sharing has taken on a more abstract communicative dimension. Here, sharing is a category of speech, a type of talk, characterized by the qualities of openness and honesty, and commonly associated with the values and virtues of trust, reciprocity, equality, and intimacy, among others. This is the "sharing" for which AA members are thanked after telling the group about their struggles with the bottle; it is the "sharing" referred to by Beck and Beck-Gernsheim (1995) in their characterization of the modern, or pure, relationship as based on "a couple sharing emotions," and as no longer based on a family "sharing the work" (48); it is the "sharing" described by Donal Carbaugh (1988) in his groundbreaking work on *Donahue* in the mid-1980s. This type of sharing—which clearly resonates with assumptions about the self, and how it must communicate itself and with others, that characterize our so-called therapeutic culture—has its roots in the Oxford Group, an early twentieth-century Christian group that gave birth to the pre–World War II Moral Rearmament movement and to Alcoholics Anonymous. Sharing, in the Oxford Group, was the public confession of sins or, to put it in terms that show its current relevance, the public declaration of weakness for the sake of redemption. Sharing thus understood is the communication of a deep personal truth and, in the Oxford Group, was perceived as the bedrock of man's relationship with God, but also, and perhaps primarily, with other people (especially one's spouse). Since at least Saint Augustine, religious language has drawn on the confessional tradition of expunging internal guilt, or externalizing the profane (whose Latin etymology means outside or before, *pro-*, the temple, *-fanum*)—only with the modern therapeutic tradition, however, has it been called *sharing*. To the religious leaders of the Oxford Group the psychological import of sharing was clear: sharing made you feel better; it contributed to your well-being.

Today, the category of speech we call *sharing* tends to involve the communication of emotions, and the word itself functions as what speech act theorists call a metalinguistic performative verb, or illocutionary force-indicating device—meaning that it tells us what kind of speech is about to follow. Put simply, if someone tells us he has something he would like to share with us, we will get ready to hear something of personal, emotional consequence, and not, for instance, that the toothpaste has run out. This sense of sharing—as the type of communication on which contemporary friendships and intimate relationships are based—is part of the metaphor of sharing in the context of SNSs and in its digital sense more generally.

The final sense of sharing that informs its multilayered status as a digital keyword harkens back to its distributive logic, taken up, for instance, in the neologism *the sharing economy*. The sharing economy incorporates elements of both production (think Wikipedia and Linux) and consumption (think Couchsurfing and Zipcar), and is conceived as a predominantly technological phenomenon among the tech-savvy online. An analysis of sixty-three newspaper articles about "collaborative consumption" published between May 2010 and April 2012, for instance, shows technology to be central to this phenomenon, both in *enabling* new ways of distributing goods (searchable and geo-tagged databases make it easier to locate and therefore borrow your neighbor's power drill) and, interestingly, in *driving* both new and old sharing behaviors. Here, a causal argument occasionally holds that sharing online (updating statuses, tweeting, etc.) leads people to want to share offline. Regardless of the empirical accuracy or otherwise of this claim, sharing is represented as a more moral and environmentally friendly alternative to capitalist models of production and consumption. It plays heavily on interpersonal relations, promising to introduce you to your neighbors, for instance, or to reinstate the sense of community that has been driven out by, say, the alienation supposedly typical of modern urban life. It conceptualizes sociality in terms of mutuality, openness, trust, commonality, and, to a degree, equality.

This, then, is the final component of *sharing* as a digital keyword. It is an important component, both because it relates us back to our most intuitive sense of what sharing is—distributing stuff among people in a fair way—as well as because it helps to infuse the notion of sharing with its utopianism. Indeed, whatever we may think of sharing online, we would be hard pushed to think of any kind of good society in which (offline) sharing did not feature prominently—in both its distributive and communicative senses. More than this: those who say that what we do on Facebook is "not really sharing" are trying to protect the word from the deleterious influences of commercialism.[6]

As I hope is clear by now, my objective here is not to call the tech companies out for hypocrisy, and I am not overly concerned with what is and is not considered "really sharing." In trying to understand *sharing* as a keyword, I am far more interested in the ways in which the term works as a metaphor, how it "organize[s] our thoughts and actions," as Lakoff and Johnson put it (1980, 40). When we dub our digital interactions *sharing*, this encapsulates the promise of our generation's technologies, just as the telegraph, the radio, and the television came with their promises too. The promise of sharing is (at least) twofold. On the one hand, there is the promise of honest and open (computer-mediated) communication between individuals; the promise of knowledge of the self and of the other based on the verbalization of our inner thoughts and feelings. On the other hand, there is the promise of improving what many hold to be an unjust state of affairs in the realms of both production and consumption; the promise of an end to alienation, exploitation, self-centered greed, and breathtaking wastefulness. For some the incongruity is too much, but these are the associations with *sharing* that tech companies seek to be identified with, and that utopianists—the vast majority of whom do not have a stake in the success or failure of this or that platform—believe will be realized as a result of the deeper embedding of social media in our everyday lives. For them, the internet really does promise improved sociability and really is the technological key to a better society through the spread of the (technology-driven) sharing economy. The concept of sharing represents both a set of values and a set of practices such that the latter, it is claimed, will help us achieve the former. As a keyword for the digital age, *sharing* bears the promise for a better society, while requiring us always to keep in mind the political economy of the structures—digital and otherwise—through which we carry out our various practices of sharing.

See in this volume: cloud, community, culture, democracy, digital, hacker, internet, meme, personalization

See in Williams: capitalism, communication, community, labour, media, socialism, technology

Notes

1 At least for English speakers. The question of the cultural associations triggered by the word *sharing* in other languages is a fascinating one that has potential to shed much light on the situatedness of our conceptions of the promises and contradictions of the digital era.
2 The methodology for and detailed results from this research can be found in John 2013a.
3 Even this sense of sharing may be metaphorical, as perhaps the earliest use of the word was to refer to the groin, where the trunk of the body divides into two legs.
4 This is described superbly in Tamar Katriel's ethnography of how Israeli children share candies (Katriel 1987).
5 For brevity's sake, I am eliding two senses of *sharing*—sharing as dividing, and sharing as having in common. Both of these senses fall within what I call the distributive logic of sharing (for more, see John 2013b).
6 I do not have the space here to explore the idea that the high-tech entrepreneurs behind today's tech behemoths also feel a commitment to an ethos of sharing (normatively understood). If I did I would discuss Yuval Dror's work on how some high-tech companies (including Facebook) claim that

they are not in it for the money (Dror 2013) and Fred Turner's exploration of the link between Google and the antiestablishmentarianism of the Burning Man festival (Turner 2009). See also Barbrook and Cameron's polemic against the "Californian ideology" (Barbrook and Cameron 1996).

References

Barbrook, R., and A. Cameron. 1996. "The Californian Ideology." *Science as Culture* 6(1): 44–72.

Beck, U., and E. Beck-Gernsheim. 1995. *The Normal Chaos of Love*. Cambridge: Polity Press.

Carbaugh, D. A. 1988. *Talking American: Cultural Discourses on Donahue*. Norwood, NJ: Ablex Pub. Corp.

Dror, Y. 2013. "'We are not here for the money': Founders' Manifestos." *New Media & Society*. doi: 10.1177/1461444813506974.

John, N. A. 2013a. "Sharing and Web 2.0: The Emergence of a Keyword." *New Media & Society* 15(2): 167–82. doi: 10.1177/1461444812450684.

———. 2013b. "The Social Logics of Sharing." *Communication Review* 16(3): 113–31. doi: 10.1080/10714421.2013.807119.

Katriel, T. 1987. "'Bexibùdim!': Ritualized Sharing among Israeli Children." *Language in Society* 16(03): 305–20.

Lakoff, G., and M. Johnson. 1980. *Metaphors We Live By*. Chicago: University of Chicago Press.

Turner, F. 2009. "Burning Man at Google: A Cultural Infrastructure for New Media Production." *New Media & Society* 11(1–2): 73–94.

25

Surrogate

Jeffrey Drouin

Historical scholarship in literary studies is increasingly dependent upon digital objects that stand as substitutes for printed or manuscript material. Digital surrogates often mimic the functionalities of codices and other material formats, ostensibly to reproduce the experience of handling the originals while taking advantage of the vastly different cognitive and representational possibilities afforded by the new medium. There has been much theorizing since McLuhan (1964) on how new media contain the old: indeed we might say scholars involved in historicist criticism are increasingly making digital simulacra into effigies of print. Archives of digitized print materials do not pretend to replace the experience of the original but nonetheless promise, implicitly if not explicitly, a way of engaging with the attributes of those objects in order to facilitate scholarly judgments about them. Thus digitized editions embody the ecclesiastical origins of the surrogate—a term derived from the Latin for "[a] person appointed by authority to act in place of another; a deputy," or one who is asked or elected (*-rogate*) over (*super+*) another—and its related concepts that impinge upon scholarly and institutional authority. When office, conceived as a symbol of power, is transferred to the realm of text, a digital edition that duplicates a print or manuscript document comes to both embody as well as symbolize the power inherent in the original it stands for. In that way, the digital surrogate is an effigy: the image of an original simultaneously worshipped and desecrated in the act of interpretation.

A digital edition is a surrogate in that it stands in for a print original (at least to the degree printed texts may be considered originals). We gain many practical benefits from using digital surrogates in literary scholarship, ranging from protection of the fragile original when a copy would suffice, to increasing access to rare materials, and rendering documents searchable and interoperable with other networked resources. Libraries have been major proponents of digital surrogates, which have long been touted by digital humanists, archivists, and special collections departments for their scholarly and preservational utility. Digital surrogates have also become levelers of class inequalities among researchers, allowing access for those who cannot afford to travel to the archives that house rare artifacts. As digital humanities has flourished over the past decade or so, the searchability and interoperability of digital texts through the TEI encoding guidelines and Dublin Core metadata standards have expanded the usefulness of digital surrogates for making large gestures about literary history, especially when corpus analysis can process "big data" sets, which are much larger than can be processed by scholars individually or in the aggregate. There is no denying the innovative possibilities in massive digital corpora.

However, when the move toward corpus-level analysis leads to inferences about texts in the aggregate, we necessarily ignore to a large degree the individual works that make up the corpus. A number of epistemological and practical questions follow: Each work says *something* from a particular point of view, so how can we be sure that our corpus-level inferences are accurate or meaningful? Is the singular text lost in the move toward searchable corpora? Is it possible to develop a methodology that synthesizes search-based queries and the uniqueness of the underlying texts? When we use digital methods upon digitized texts, what exactly is the object of study—the semantic unit, the text, the genre, the written language? And, if we attempt to compensate for the blind spots of large-scale analysis by selecting individual works from the digital corpus, how can we know whether we are adequately filling in the gaps?

While a digital edition offers built-in functionalities and research possibilities unavailable in a printed object, the interface also erases many physical traits of the original, such as size, weight, paper quality, and ink saturation—all of which are crucial in matters of historical and bibliographic analysis. Take the central example of this essay: the Modernist Journals Project (MJP) (http://modjourn.org) features an edition of *BLAST*, an important avant-garde magazine from one hundred years ago known for its radical experiments in typography and poetics.[1] Even though the MJP offers high-fidelity scans of the original pages, the physical impact of the magazine is lost in translation. A brief description of the original format may help: the bibliographic information supplied on the landing page of the digital edition indicates that the 212 pages of the first issue (June 1914) are 30.5 cm long and 24.8 cm wide (more than 12 inches and 10 inches, respectively). A reader could use a ruler or tape measure as a visual aid in comprehending the size, since the surrogate will almost certainly appear smaller on a screen. Yet in no way does the comprehension of measurements equal the aesthetic apprehension of seeing—and holding and smelling—a codex that is roughly the area of a small poster, that is twice as wide when opened up, and whose thick paper renders it roughly 6.35 cm (2.5 inches) deep, weighing around 1 kg (2.25 lbs.), and supporting the heavily saturated block letters that often stand over 2.5 cm (1 inch) tall on the page as if they are autonomous objects.

The physical experience of reading *BLAST* necessarily contributes to the interpretation of its content, since such a solid, impactful object is diametrically opposed to the ephemerality normally expected of magazines: it is a Vorticist manifesto attempting to break art and literature aesthetically, morally, and physically—to "be an avenue for all those vivid and violent ideas that could reach the Public in no other way" by bringing "to the surface a laugh like a bomb" (Lewis, "Long Live" 7, "MANIFESTO" 31).

Indeed, the kinetic typography that often spans juxtaposed pages produces a visual effect whose immensity corroborates its revolutionary assertions. The accompanying image presents a digital imitation of two juxtaposed pages from *BLAST* that demonstrate the interplay of typography and ideology. The series of "Blasts" and "Blesses" comprising this section of the manifesto take aim at the passé while asserting an English art that is nationalist and industrial in temper. Throughout most of modern history, English artists and writers looked enviously upon their French colleagues as being more advanced. Here, however, the attacks upon French culture by the magazine's "Primitive Mercenaries in the Modern World" ("MANIFESTO" 30) seek to create a new space for English art that far surpasses its rival. A key tactic in surpassing the French is to embrace the opposing energies of an explosion: "We fight first on

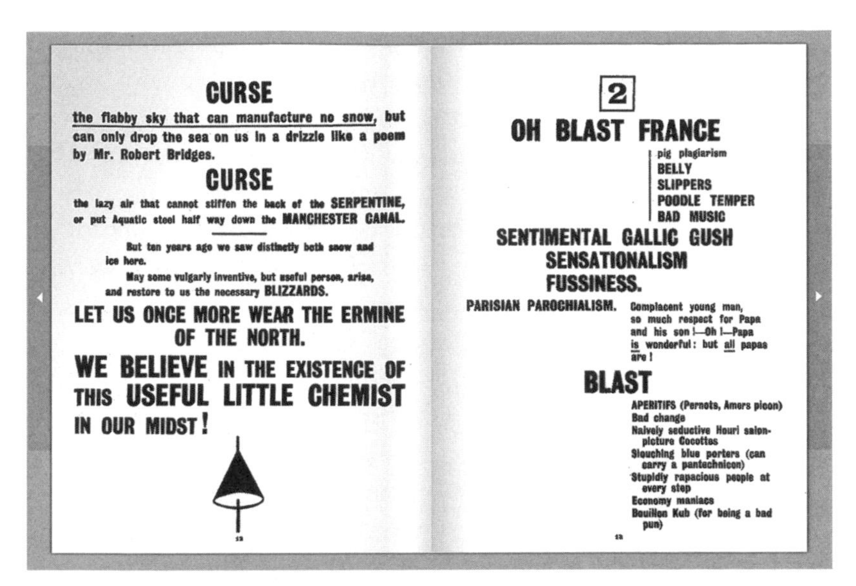

CURSE

the flabby sky that can manufacture no snow, but can only drop the sea on us in a drizzle like a poem by Mr. Robert Bridges.

CURSE

the lazy air that cannot stiffen the back of the SERPENTINE, or put Aquatic steel half way down the MANCHESTER CANAL.

But ten years ago we saw distinctly both snow and ice here.

May some vulgarly inventive, but useful person, arise, and restore to us the necessary BLIZZARDS.

LET US ONCE MORE WEAR THE ERMINE OF THE NORTH.

WE BELIEVE IN THE EXISTENCE OF THIS USEFUL LITTLE CHEMIST IN OUR MIDST!

2

OH BLAST FRANCE

pig plagiarism
BELLY
SLIPPERS
POODLE TEMPER
BAD MUSIC

SENTIMENTAL GALLIC GUSH
SENSATIONALISM
FUSSINESS.

PARISIAN PAROCHIALISM. Complacent young man, so much respect for Papa and his son!—Oh!—Papa is wonderful: but all papas are!

BLAST

APERITIFS (Pernots, Amers picon)
Bad change
Naively seductive Houri salon-picture Cocottes
Slouching blue porters (can carry a pantechnicon)
Stupidly rapacious people at every step
Economy maniacs
Bouillon Kub (for being a bad pun)

Two pages of "Manifesto" from *BLAST*.

one side, then on the other, but always for the SAME cause, which is neither side or both sides and ours" (30). Hence Vorticism, taking its cue from the vortex or whirlpool (as well as adolescence), deliberately embodies opposing forces at their point of maximum concentration, which is simultaneously their point of cancellation. These self-contradictory position statements explain the typographical interplay of absence and presence on the magazine's pages, where bullet lists occupy the left or right side of a page, while a vertical line divides text on one side of the page from a blank space on the other (taking a cue from commercial advertising), or where there seems to be a diagonal line separating absence and presence across two juxtaposed pages, as in the screenshot reproduced here. The reader can fully appreciate the amount of energy required to embody these principles only while situated before the text arrayed across an area of 1,513 square centimeters (240 square inches) plunked solidly upon a table.

The reason the image reproduced in this essay is currently not part of the MJP, which I codirect, may deserve a personal anecdotal description. In my quest to view digital pages of *BLAST* juxtaposed as they are in the original, I discovered it was not possible to view both pages at once, a problem I sought to resolve by first submitting a PDF version of the magazine to FlipSnack[2] so that I could observe it onscreen (or at least the first sixteen pages available without my paying for the service) and then embedding it on a teaching blog linked below.[3] I also fiddled with the Two Page View option (with the selected option to Show Cover Page) in Adobe Acrobat, which works well on a computer but cannot be embedded on a website. A codex-simulating view option will be available on the MJP website in the coming year, but, more interesting, in the process of trying to view the original print format digitally, I have found myself jerry-rigging a digital simulacrum of it.

The *Oxford English Dictionary* informs us that a simulacrum is a "material image, made as a representation of some deity, person, or thing"; it possesses "merely the form or appearance of a certain thing, without possessing its substance or proper qualities"; it is "a mere image, a specious imitation or likeness, of something." In other words, it fulfills the role of surrogate as a substitute deputed by authority yet lacks the true substance of that for which it stands. The association of a simulacrum with a deity—and inadequacy—seems apt in light of the digital *BLAST* as well as electronic editing and scholarship in general. A goal of the simulacrum is to return to some originary state, "to see the thing as in itself it really is" or was, to paraphrase Matthew Arnold. But in translating *BLAST* into the new medium, which cannot adequately duplicate the physical attributes inherent in its meaning, are we not in fact moving away from the original?

Every act of copying creates an effigy: a likeness, portrait, or image that lacks the character of the original yet stands in for our pursuit of it. Like the other terms in this conceptual cluster—*surrogate* and *simulacrum*—*effigy* bears the undertones of a symbol of something holy to be revered, as well as the substitute for something profane to be desecrated. It is telling that the various definitions of *effigy* draw upon ecclesiastical and judicial terminology. In that light, my hasty decision to feed *BLAST* to FlipSnack *in effigy* betrays an attempt to incarcerate and subject the original to tools under my command: "fig. . . . to inflict upon an image the semblance of the punishment which the original is considered to have deserved; formerly done by way of carrying out a judicial sentence on a criminal who had escaped." In this case, we must ask what are the processes of historicity that motivate the raising of this effigy.

The digital surrogate clearly has its impetus in preserving what Walter Benjamin describes as the aura, the "presence of the original" that is "the prerequisite to the concept of authenticity" (220). For Benjamin, the aura belongs to a unique work of art, such as a painting or sculpture, whose authenticity thereby exudes "the essence of all that is transmissible from its beginning, ranging from its substantive duration to its testimony to the history which it has experienced" (221). Digital surrogates like that of *BLAST* depend upon something like the aura for their attraction as objects of research, even though the nature of that aura, if it may be so called, is complicated by the fact that *the original* was a born-print magazine, not a unique manual creation. Here the judicial valence of the surrogate becomes relevant in establishing legitimacy. According to Benjamin, before modern machinery, manual reproduction was "usually branded as a forgery," since "the original preserved all its own authority" (220). Likewise, a mechanically reproduced copy of an artwork lacks the authenticity of the original because its "substantive duration ceases to matter," which jeopardizes its historical testimony (221). The stoic value of "substantive duration" appears to ascribe selfhood, to venerate the individuality of the artwork as it ages and continues to assert its voice over time and through changing circumstances—in other words, it appears that the digital surrogate risks losing precisely what we are attempting to recover through the study of original texts. Still there is something to be gained: for a magazine like *BLAST*, however innovative it was, its centripetal pull derives from its multiple birth in that fateful summer of 1914. Its aura has never been the same kind as that ascribed to unique masterpieces such as the *Mona Lisa* or Picasso's *Still Life with Chair Caning*; as a text, the work makes up a complex and still-unfolding archive comprising all of its variants, including *avant-texte*, print copies, digital surrogates, and derivative data. The work transcends its various instantiations, which are now augmented through the possibilities of the digital medium.

Benjamin's assertion that technical reproduction does not constitute a forgery offers an interesting twist: a mechanically or digitally reproduced artwork comes into being not through an individual's mimicking the technique or vision of the original artist but through a technical process independent of and removed from the original, such as what takes place in a photography lab or on the cutting floor of the sound film studio. The photographic negative might constitute an original in some sense, but the possibilities it offers for postproduction manipulation are manifold, including very high levels of magnification in order to reveal elements of the subject invisible to the naked eye, or, in the case of cinema, to study motion through manipulation of projection speed and direction. The high-definition mimesis of digital scanning provides many more photographic possibilities. Print magazines are in this way similar to photo negatives, especially after they are translated and transcribed into a digital medium; they become metamorphosed into a different kind of art—no longer a printed text. Like a pagan animal deity, which is less than human insofar as it embodies a single characteristic, yet thereby far more powerful as a force of nature, the digital surrogate diminishes the aura of the printed original but still wields an enchantment all its own.

Lest this meditation be misconstrued as a kind of Proustian obsession with "The Sweet Cheat Gone," it must be acknowledged that it is illusory to demand total knowledge of our Albertine. It is perhaps fitting that the hunt for the fugitive original now extends to mechanically reproduced print "originals" that, when read through their digital surrogates, do just the same. In what ways do they evade the reader—or, now, the computational investigator? Is the data output of the digital surrogate fundamentally separate from the object from which it derives? Our response will depend on the language we use to inquire after digital materiality, which remains central to the current information age, as well as other essays in this volume (see **analog**, **cloud**, **digital**, etc.; Sterne; Kirschenbaum, et al.). As for this essay, the digital surrogate-*cum*-effigy seems to approach the character of the fetish: "a means of enchantment . . . or superstitious dread"; "an inanimate object worshipped by preliterate peoples on account of its supposed inherent magical powers, or as being animated by a spirit"; "something irrationally reverenced." Perhaps, taking another cue from Benjamin, we still read the digital surrogate as a hieroglyph—not just on the level of the linguistic or even bibliographic code, but in some other mode of apprehension for which we have not yet developed the pictorial language.

See in this volume: analog, archive, cloud, digital, memory, mirror, prototype

See in Williams: aesthetic, creative, fiction, image, literature, nature, representative, subjective

Notes

1 http://modjourn.org/render.php?id=1158591480633184&view=mjp_object.
2 http://flipsnack.com.
3 *BLAST*, no. 1 (June 1914), http://www.flipsnack.com/7C66E8EC5A8/fvc8bf5j.html; *BLAST* no. 2 (July 1915, "War Number"), http://www.flipsnack.com/7C66E8EC5A8/fxcfol89.html.

References

Benjamin, Walter. "The Work of Art in the Age of Mechanical Reproduction." In *Illuminations: Essays and Reflections*, edited by Hannah Arendt, translated by Harry Zohn, 217–51. New York: Schocken Books, 1969.

Kirschenbaum, Matthew, et al. "Digital Materiality: Preserving Access to Computers as Complete Environments." iPres 2009 (October 2009). http://escholarship.org/uc/item/7d3465vg.

Lewis, Wyndham. "Long Live the Vortex!" *BLAST* 1(1) (June 1914): 7–8.

———. "MANIFESTO." *BLAST* 1(1) (June 1914): 30–43.

McLuhan, Marshall. *Understanding Media: The Extensions of Man*. London: Routledge and Kegan Paul, 1964.

Sterne, Jonathan. "What Do We Want? Materiality! When Do We Want It? Now!" In *Media Technologies: Essays on Communication, Materiality and Society*, edited by Tarleton Gillespie, Kirsten Foote, and Pablo Boczkowski, 119–28. Cambridge, MA: MIT Press.

Appendix: Over Two Hundred Digital Keywords

(**bolded** words are included in this volume)

access
activism
alert
algorithm
analog
analy(sis/tics)
anonym-
app (aratus/lication)
architecture
archive
art
artificial
avatar

banner
big data
bio-
blog
body
book
-bot
(broad/ narrow/ pod)-cast
buffer
burn

cache
capital
celebrity
class
click(bait)
cloud
code
collaborat(e/ion)
collective
comment
communicat(e/ion)
community
compute
content
control
cookie
cooperat-
copy
crawl
create
crowd(sourcing)
crypt
culture
currency
cyber-

data
database
delete
democracy
design

digit(al/ization)
domain
drone

economy
ecosystem
e-(lectronic)
elite
emotion
error
event

fair
feed
filter
flame
flow
follow
form(at)
forum
frag
free
friend
future

gam(e/er/ing)
geek
gender
globe
graph

hack(er/ing)
hashtag
history
hit
human
humanities

hybrid
hyper-

i-
icon
identi(fy/ty)
image
index
industry
information
infrastructure
innovation
intelligence
interact
interface
internet
interpret

keyword
knowledge

labor
language
law
liking
link
list
local
-log
logistics

machine
make(r)
markup
mass
material
media

meh
meme
memory
message
meta-
mirror
model
modern
mouse

narrative
narrow
native
nature
network
new
noisc

object
open
opinion
opt
organ-(ic/izational)
origin

packet
page
participation
password
path
peer
person-
(a/alization)
ping
platform
play(er)
plug
pod
politics
press
privacy
process
profile
program
protocol
prototype
proxy

query
queue

rank
real-time
resolution
revolution
right
risk
robot
route

scale
scrape
screen
script
search
secure
self(ie)
senses
serv(e/er/ice)
sex
sharing
sign(al)
simulation
site

smart
snap
soci(al/ety)
sort
sound
space
spam
spectacle
spy
stick
stream
structure
submission
surf(ace)
surrogate
surveil(lance)
system

tag
target(ing)
technology
terms
text
theory
thing
time
token
track
traffic
trans-(coding/duction/fer/mission/lation/pond/etc.)trending
trigger
troll
tweet

user

value
vector
view
virus (bug/crawler/spider/web/worm/etc.)
vlog

wall
-ware (hard/soft/shelf/wet, etc.)
web
widget
wiki
wireless
work
write

you

zoom

About the Contributors

Rosemary Avance received her PhD from the Annenberg School for Communication at the University of Pennsylvania. Her research and teaching center on the intersection of new media, religion, and modernity. Her work site is http://www.rosemaryavance.com/.

Saugata Bhaduri is professor of English and associate dean of students at Jawaharlal Nehru University, New Delhi, India. His areas of research and teaching interest include popular culture, of both folk and the mass-mediatized sort. See http://www.jnu.ac.in/Faculty/bhaduri/cv.pdf for more.

Sandra Braman is John Paul Abbott Professor of Liberal Arts and professor of communication at Texas A&M University with research and teaching interests in digital technologies and their policy implications. Her work site is people.tamu.edu/~braman/.

Gabriella Coleman is the Wolfe Chair in Scientific and Technological Literacy at McGill University with teaching interests in computer hackers and digital activism. She has authored two books on computer hackers. For more, see http://gabriellacoleman.org/.

Jeffrey Drouin is assistant professor of English and codirector of the Modernist Journals Project at the University of Tulsa. He is the author of *James Joyce, Science, and Modernist Print Culture: "The Einstein of English Fiction"* (Routledge, 2014) and creator of *Ecclesiastical Proust Archive*, http://proustarchive.org.

Christina Dunbar-Hester teaches courses on technology and culture in the Annenberg School for Communication and Journalism at the University of Southern California, where she works as an assistant professor. She is the author of *Low Power to the People: Pirates, Protest, and Politics in FM Radio Activism* (MIT Press, 2014). She writes about media activism and political engagement with technology, and many of her pieces can be found on http://usc.academia.edu/ChristinaDunbarHester.

Adam Fish is a social anthropologist of digital industries and digital activism who teaches in the Sociology Department at Lancaster University in the United Kingdom. His work site is http://www.lancaster.ac.uk/arts-and-social-sciences/about-us/people/adam-fish.

Hope Forsyth is a JD candidate with research interests in copyright and media history at the University of Tulsa, where she also earned her honors bachelors degree in communication (with minors in English and philosophy) as a Presidential Scholar.

Bernard Geoghegan is assistant professor at the Institut für Kulturwissenschaft at the Humboldt-Universität zu Berlin, where he teaches courses on media theory and the history of technology. See http://bernardg.com/ for more.

Tarleton Gillespie is a principal researcher at Microsoft Research, New England, and an associate professor in the Department of Communication at Cornell University. He is the author of *Wired Shut: Copyright and the Shape of Digital Culture* (MIT Press, 2007) and the cofounder of the scholarly blog *Culture Digitally*, http://culturedigitally.org/.

Katherine D. Harris is associate professor in English at San Jose State University, where she teaches about topics in literature and technology ranging from the mechanization of the printing press in nineteenth-century England to current uses of narrative in gaming. Find her latest work and public lecture schedule at http://triproftri.wordpress.com.

Nicholas A. John is lecturer in the Department of Communication and Journalism at the Hebrew University of Jerusalem. His research interests include technology and society, social media, and sharing. Find him at http://nicholasjohn.huji.ac.il.

Christopher Kelty is professor in the Institute of Society and Genetics, with appointments in the Department of Information Sciences and the Department of Anthropology at the University of California, Los Angeles. The author of *Two Bits: The Cultural Significance of Free Software* (Duke University Press, 2008), he teaches courses on the history of modern thought, science studies, and anthropology. More is available at http://kelty.org/.

Rasmus Kleis Nielsen is director of research at the Reuters Institute for the Study of Journalism, University of Oxford and editor of the *International Journal of Press/Politics*. His research deals with political communication, changes in news media and journalism around the world, and the role of digital technology in these areas. More on his work here: http://rasmuskleisnielsen.net/.

Benjamin Peters is assistant professor of communication at the University of Tulsa, where he teaches courses on media history and theory with a particular emphasis on information technologies. He keeps working notes at http://petersbenjamin.wordpress.com.

John Durham Peters is A. Craig Baird Professor in Communication Studies at the University of Iowa, where he teaches courses on the cultural history of media and social theory. His work site is http://johndurhampeters.wordpress.com.

Steven Schrag is a PhD student in the Annenberg School of Communication at the University of Pennsylvania, with research interests at the intersection of technology, worldbuilding, and memory. For more information: https://www.asc.upenn.edu/people/students/steven-schrag.

Stephanie Ricker Schulte is associate chair and associate professor of communication at the University of Arkansas, where she researches communication technologies, popular culture, and transnational media policy. She is the author of *Cached: Decoding the Internet in Global Popular Culture* (New York University Press, 2013).

Limor Shifman is associate professor in the Department of Communication and Journalism at the Hebrew University of Jerusalem. Her research and teaching interests include the social construction of humor, popular culture, and new media. For more information: http://pluto.huji.ac.il/~mslimors/.

Julia Sonnevend is assistant professor of communication studies at the University of Michigan, where she teaches courses on events and symbols, icons and performances in global media. For more information: http://julia-sonnevend.com.

Jonathan Sterne is professor and James McGill Chair in Culture and Technology at McGill University. He writes and teaches on sound, media theory and history, technology and culture, and disability. His latest book is *MP3: The Meaning of a Format* (Duke University Press, 2012). For more, see http://sterneworks.org.

Thomas Streeter is professor of sociology at the University of Vermont, where he teaches about media and culture, while researching the intersections between law, technology, culture, and language. He is currently studying the effects of the ongoing digitization of legal documentation on legal practices. More about him can be found at http://www.uvm.edu/~tstreete.

Ted Striphas is associate professor of communication at the University of Colorado–Boulder. He teaches courses on the history and philosophy of technology; on the politics of everyday life; and on cultural studies and keywords. He is author of *The Late Age of Print* (Columbia University Press, 2009) and is at work on his next book, *Algorithmic Culture.* Twitter: @striphas.

Fred Turner is professor in the Department of Communication at Stanford University, where he teaches on the intersection of media, technology, and American cultural history. For more, see http://fredturner.stanford.edu/.

Guobin Yang is associate professor of communication and sociology at the Annenberg School for Communication and the Department of Sociology at the University of Pennsylvania, where he writes and teaches on digital media, political communication, and social movements in global and Chinese contexts. See https://www.asc.upenn.edu/people/faculty/guobin-yang-phd.

Index